Advances in Underwater Technology, Ocean Science and Offshore Engineering

Volume 30

Subsea International '93

Low Cost Subsea Production Systems

ADVANCES IN UNDERWATER TECHNOLOGY, OCEAN SCIENCE AND OFFSHORE ENGINEERING

Vol. 1. Developments in Diving Technology
Vol. 2. Design and Installation of Subsea Systems
Vol. 3. Offshore Site Investigation
Vol. 4. Evaluation, Comparison and Calibration of Oceanographic Instruments
Vol. 5. Submersible Technology
Vol. 6. Oceanology
Vol. 7. Subsea Control and Data Acquisition
Vol. 8. Exclusive Economic Zones
Vol. 9. Stationing and Stability of Semi-submersibles
Vol. 10. Modular Subsea Production Systems
Vol. 11. Underwater Construction: Development and Potential
Vol. 12. Modelling the Offshore Environment
Vol. 13. Economics of Floating Production Systems
Vol. 14. Submersible Technology: Adapting to Change
Vol. 15. Technology Common to Aero and Marine Engineering
Vol. 16. Oceanology '88
Vol. 17. Energy for Islands
Vol. 18. Disposal of Radioactive Waste in Subsea Sediments
Vol. 19. Diverless and Deepwater Technology
Vol. 20. Subsea International '89: Second Generation Subsea Production Systems
Vol. 21. NDT: Advances in Underwater Inspection Methods
Vol. 22. Subsea Control and Data Acquisition: Technology and Experience
Vol. 23. Subtech '89. Fitness for Purpose
Vol. 24. Advances in Subsea Pipeline Engineering and Technology
Vol. 25. Safety in Offshore Drilling. The Role of Shallow Gas Surveys
Vol. 26. Environmental Forces on Offshore Structures and their Prediction
Vol. 27. Subtech '91. Back to the Future
Vol. 28. Offshore Site Investigation and Foundation Behaviour
Vol. 29. Wave Kinematics and Environmental Forces
Vol. 30. Subsea International '93: Low Cost Subsea Production Systems

CONFERENCE PLANNING COMMITTEE

Advances in
Underwater Technology,
Ocean Science and
Offshore Engineering

Volume 30

Subsea International '93

Low Cost Subsea Production Systems

Papers presented at a conference
organized by the Society for Underwater Technology
and held in Aberdeen, UK, April 28–29, 1993.

SPRINGER-SCIENCE+BUSINESS MEDIA, B.V.

Library of Congress Cataloging-in-Publication Data

Subsea International '93 (1993 : Aberdeen, Scotland)
Subsea International '93: low cost subsea production systems / papers presented at a conference sponsored by the Society for Underwater Technology and held in Aberdeen, UK, April 23-29, 1993.
p. cm. -- (Advances in underwater technology, ocean science, and offshore engineering ; v. 30)
At head of title: International conference.
ISBN 978-0-7923-2243-6 ISBN 978-94-011-1717-3 (eBook)
DOI 10.1007/978-94-011-1717-3
1. Oil well drilling, Submarine--Costs--Congresses. 2. Oil well drilling, Submarine--Cost effectiveness--Congresses. 3. Oil well drilling, Submarine--Technological innovations--Congresses. I. Society for Underwater Technology. II. Title. III. Title: Low cost subsea production systems. IV. Series.
TN871.3.S898 1993
622'.33819--dc20 93-18651

Printed on acid-free paper

Contents

Society for Underwater Technology

The Society was founded in 1966 to promote the further understanding of the underwater environment. It is a multi-disciplinary body with a worldwide membership of scientists and engineers who are active or have a common interest in underwater technology, ocean science and offshore engineering.

Committees

The Society has a number of Committees to study such topics as:

- Diving and Submersibles
- Offshore Site Investigation and Geotechnics
- Environmental Forces and Physical Oceanography
- Ocean Resources
- Subsea Engineering and Operations
- Education and Training

Conference and Seminars

An extensive programme is organized to cater for the diverse interests and needs of the membership. An annual programme usually comprises four conferences and a much greater number of one-day seminars plus evening meetings and an occasional visit to a place of technical interest. The Society has organized over 100 seminars in London, Aberdeen and other appropriate centres during the past decade. Attendance at these events is available at significantly reduced levels of registration fees for Members or staff of Corporate Members.

Publications

Proceedings of the more recent conferences have been published in this series of *Advances in Underwater Technology, Ocean Science and Offshore Engineering.* These and other publications produced separately by the Society are available through the Society to members at a reduced cost. A careers pack 'Oceans of Opportunity' has been produced by the Society in response to the growing demand by students schools and colleges for up-to-date information.

Journal

The Society's quarterly journal *Underwater Technology* caters for the whole spectrum of the inter-disciplinary interests and professional involvement of its readership. It includes papers from authoritative international sources on such subjects as:

- Diving Technology and Physiology
- Civil Engineering
- Submersible Design and Operation
- Geology and Geophysics

Subsea Systems
Naval Architecture
Marine Biology and Pollution
Oceanography
Petroleum Exploration and Production
Environmental Data

An Editorial Board has responsibility for ensuring that a high standard of quality and presentation of papers reflects a coherent and balanced coverage of the Society's diverse subject interests; through the Editorial Board, a procedure for assessment of papers is conducted.

Endowment fund

A separate fund has been established to provide tangible incentives to students to acquire knowledge and skills in underwater technology or related aspects of ocean science and offshore engineering. Postgraduate students have been sponsored to study to MSc level and subject to the growth of the fund it is hoped to extend this activity.

Awards

An annual President's Award is presented for a major achievement in underwater technology. In addition there is a series of sponsored annual awards by some Corporate Members for the best contribution to diving operations and oceanography, and for the best technical paper in the Journal

FURTHER INFORMATION

If you would like to receive further details, please contact
Society for Underwater Technology, The Memorial Building, 76 Mark Lane,
London EC3R 7JN.
Telephone: 071-481 0750; Telex: 886481 I Mar E G; Fax: 071-481 4001.

Session 1
Technical Refinement

LOW COST AUTONOMOUS SUBSEA WATER INJECTION SYSTEMS

W D LOTH
W D Loth & Co Ltd
Stammerham Business Centre
Capel Road, Rusper,
West Sussex, RH12 4PZ
United Kingdom

C A WALKER
Marconi Underwater Systems
Elletra Avenue
Waterlooville, Hants
PO7 7X9
United Kingdom

ABSTRACT Autonomous systems hold promise for reduced cost and more environmentally friendly subsea control systems. A Long Range Remotely Operated Vehicle is described which provides a simple control system option based on extensions of field proven technology.

INTRODUCTION

In order to reach full potential, autonomous subsea well control systems must be developed for producing wells. Nevertheless, consideration of a water injection well forms a useful starting point and the concepts described in this paper could form the basis of an autonomous control system for multiple well subsea developments. The incentive for autonomous operations are cost savings and environmental advantages stemming from elimination of the control umbilical. The limited telemetry utilized in the system described will be accomplished with acoustic signals transmitted along the flowline.

Although it may seem premature to be considering improvements in autonomous systems, when such systems have yet to reach the point of serious commercial application, there is no better time to explore a wide range of possibilities than in the formative days of new technology. The concept described leads to major simplification and even lower cost autonomous systems by eliminating local power generation and many control system components. Although the concept may, at first glance, appear to be somewhat futuristic it is simple, straightforward, and based on technology well proven in underwater defence operations.

OPERATION OF THE AUTONOMOUS SYSTEM.

A conventional subsea tree, which could be equipped with standard valve actuators, is fitted with a local monitoring system. For a water injection well this

Volume 30: Subsea International '93, 3–12.

would include wellhead pressures and simple electronics to monitor a "minder" signal from the surface facility. Output of pressure transducers, and the input from the "minder" signal, are routed to a microprocessor coupled to a single, battery powered, pilot for a single hydraulic vent valve.

As long as flowline and annulus pressures remain within limits, and the "minder" signal is received, the tree valves are held open. When these conditions are not met, the pilot operated vent is released and all valves fail closed. There is no routine telemetry to the surface but a single signal is returned to confirm that the tree is shut in. The microprocessor could also be coupled with a small logging device to periodically store readings of wellhead pressures.

When the surface facility records a signal that the well has been shut in, an action initiated from the surface facility in most instances, and a decision has been taken to re-open the well, a Long Range Remotely Operated Vehicle (LRROV) is prepared and launched. The LRROV swims to the designated tree.

When the LRROV approaches the tree, the vehicle identifies the docking point and mechanically couples itself to establish electrical and hydraulic communication. This is a totally remote, self-controlled, operation although the status of manoeuvres will be monitored remotely. Once the LRROV is docked, the initial activity is to interrogate the data logging device to confirm the reason for shut-in, and current pressure status, to ensure that it is safe to open the tree valves. Sufficient logic could be readily carried within the LRROV so that the valves could be reopened on predetermined acceptance criteria or a report could be returned to the surface facility so that a decision could be made remotely.

When it is safe to recommence operations, valves are reopened by injecting a water/glycol mix into a single stab point. The vent/pilot system is reset, valves are reopened, and an accumulator recharged to accommodate small amounts of actuator leakage. Seawater powered actuators could be substituted for the conventional units but a decision as to which system is more cost effective would require more detailed appraisal. It would be also possible at this time to recharge the small battery required for the microprocessor and holding power for the pilot valve. After clearing the data logger, the LRROV swims back to the surface facility where it is recovered and made ready for the next mission.

DESCRIPTION OF COMPONENTS.

Subsea Tree

The tree, and actuated components, are standard items. Instrumentation would initially be limited to pressure readings which would be recorded only on an infrequent basis since there is no routine reporting back to the surface facility. In the most simple form, all hydraulically actuated components would be connected to a common supply/vent system. If it was required to open valves sequentially, a less common requirement for injection than for producing wells, a series of time delay flow restrictions, would be utilized. For the water injection well, a deep set high pressure safety valve may not be required but a local pressure intensifier could be utilized should it be necessary to provide higher pressure for the downhole safety valve in producing wells. There is no reason why components could not be activated through individual stabs but the additional operational flexibility may not warrant the added complexity.

Pressure data could be logged locally and recovered whenever the tree is visited by the LRROV. A primary requirement for the system is that local power consumption be minimised. This would be achieved by having a single piloted vent valve and only the most limited telemetry capability. The unit would be capable of receiving a "minder" signal periodically transmitted from the surface facility. There seems little need to reset the system, with the "minder" signal, more frequently than hourly for a water injection well but a shorter interval would probably be required for production. If at any time a signal were not transmitted from the surface, the subsea wells would automatically shut-in. The only telemetry back to the platform would be a single signal transmitted when the system is shut-in.

If necessary, additional information could be gathered. Process valving, especially electric valves, have been fitted with numerous monitoring facilities and some process valves have as many as four microprocessors with multiple input/output channels. It is not envisaged that such complex monitoring would be required in the subsea tree but it is comforting to know that the capability exists. Once shut in, the system would be entirely quiescent.

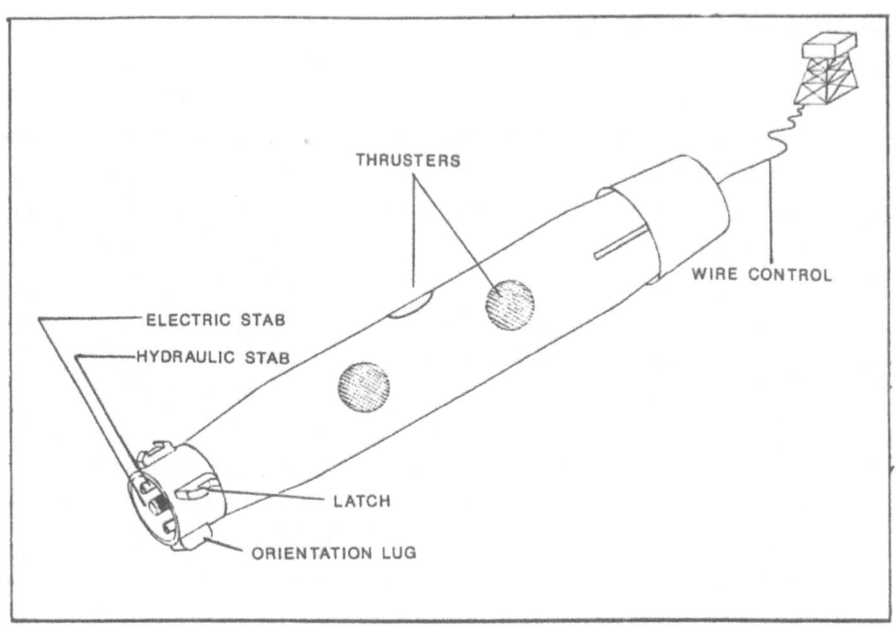

FIGURE 1.

LONG RANGE REMOTELY OPERATED VEHICLE (LRROV)

FIGURE 2.

SOURCE TECHNOLOGY – TORPEDO

It is not anticipated that the tree would incorporate elaborate or substantial docking arrangements. The nose of the LRROV would orient and be locked in a small receptacle designed only to restrain the LRROV during restart operations. The hydraulic stabs would be pressure balanced and, in the absence of a requirement for any major mechanical load resistance, the attachment need be only sufficient to resist current forces and the impact of the LRROV when docking.

LRROV

Figure 1. illustrates the concept; the vehicle remains in conceptual stage but much of the confidence in its feasibility is derived from the similarity between existing ROV operations and the use of torpedoes.

There are major differences between conventional ROV usage and the operations described in this paper. The operational range from the surface facility would be much greater than currently practical. Launch control and vehicle recovery, and the refurbishment and maintenance between missions, would require substantially less skill than for current ROVs. The vehicle would have to offer increased operational reliability and the LRROV would be a specialised vehicle with a single task application. Elements of design approach adopted for torpedoes are more appropriate than conventional ROVs. Torpedoes, Figure 2, are optimised for a single application and, in their training role, have the ability to be turned round and made available for reuse after a run with minimal effort. It is not generally appreciated that the selling price of a torpedo compares favourably with a ROV system.

Although technically possible to do so, there is no current intention to completely eliminate operator control. The system loop will include an allowance for remote command with the means to prevent the LRROV from ever performing in a manner prejudicial to safety or continued production. Since the vehicle would be utilized more frequently than units for inspection or intervention, operations have been carefully considered; it would be undesirable to require additional specialised personnel offshore. The LRROV would include the manned element of the loop at a land based location. Telemetry rates between the pilot and LRROV could be maintained in the range of 2400 to 9600 baud if standard telecommunications are utilized. It is also intended that maintenance activities on the LRROV between missions would be simplified to the point that there would be no need for specialised skills during the turn around cycle.

Although the details of vehicle design have yet to be evolved, the major subsystems have been subjected to preliminary appraisals. Final choices have yet to be made but options have been identified in all areas and preliminary selections completed. The major subsystems involved are communications, launch and recovery, navigation, docking, decision making, and hydraulic operations. By eliminating the need for a hydrodynamically inefficient umbilical and by minimizing the drag of the vehicle, it is possible to use an all electric low maintenance system. MUSL are currently testing in water their free running all electric 5 knot AUV which has a range of 300 km. For the LRROV it is envisaged that energy would be supplied by proven rechargeable battery technology and high power brushless motors would drive the propellers.

Communications are one the most unique aspects of a long range vehicle. Communications must include three links; between the pilot and the surface facility, between surface facility and the sea bed, and the link between the sea bed beneath the surface facility and the LRROV. The pilot and surface facility will normally be linked with existing communication facilities. Communications through the air/sea interface will be normally handled with a conventional umbilical to the seafloor although, for fixed installations, acoustic transmission through the jacket structure would be considered.

The use of conventional umbilicals to communicate between the local seafloor facility and the LRROV has been discounted. Umbilical size and weight would have a major impact on the LRROV if it was spooled on board the vehicle and current drag on the extended umbilical would increase the power required for transit and station holding profoundly. Acoustic communications might work in optimal conditions and could lead to significant operational performance ranges. The problem with the acoustic approach is that reliable operation cannot be guaranteed under some weather conditions or in the presence of high background noise from ships operating locally. Fibre optic umbilicals are very attractive by virtue of high bandwidth and low attenuation but the fibres can fail optically due to microbending even while appearing physically intact. The provision of external protection to provide the ruggedness necessary for oilfield handling leads to a substantial increase in cost and weight. The advantages of high bandwidth are to a large extent negated by the limitations posed by cost and weight and the requirement for more sophisticated offshore inspection, diagnostic, and maintenance capability. The favoured option is a wire guided LRROV.

The technology has been in use for over twenty years and has proven highly reliable. The data bandwidth is limited to the point that the possibility of real time video is precluded but the ruggedness of the system, low weight, and modest cost, make this the preferred choice.

The next major subsystem to be considered is launch and recovery. Three approaches have been considered. The first is chute launch and net recovery. This is a well established procedure in the defence market but has been eliminated for this application because of the design impact of making the LRROV sufficiently rugged for water entry and also potential problems associated with the communication link. The second approach would be to provide a garage on the sea bed. This solution necessitates a permanent cable being installed on the leg of the platform (the umbilical for communications could be lowered only during missions if desired). The major drawback is that the LRROV would be inaccessible between missions and additional effort is required for maintenance. This complicates the system and is not seen as an optimal approach. The selected approach is to deploy the LRROV from a garage lowered by a crane. This is an approach well established in the ROV industry and favoured for this application.

Navigation systems for the LRROV could be one of four types. The general navigation problem is relatively simple compared to that of the torpedo since the target is not moving and not taking defensive measures. The systems which might be employed are inertial guidance, pipeline following, acoustic positioning, or sonar return pattern recognition. For single satellite wells the pipeline following option would seem to be the most straightforward. Inertial systems and the sonar return pattern are favoured, over acoustic positioning, since the former two place lesser power demands at the wellhead.

Docking navigation has yet to be addressed in detail to establish an optimum solution from the cost viewpoint. Torpedo sonars are able to process returns from rapidly manoeuvring targets, and achieve good repeatability in making contact at the required point. This level of technology is not required in this application but elements of the system, combined with a mechanical configuration for final docking and orientation, provide the basis for a low risk solution.

There is relatively little decision making required. The unit must interrogate the data logger and decide whether or not the tree is safe to open. There seems little reason why this could not be accomplished on board the LRROV although, with the telemetry link provided, decision making could be confirmed or countermanded by the land based operator. With the LRROV returning to the surface facility for between mission attention, these systems could be readily recharged and there seems little requirement to develop any novel systems.

The final operation required is the manner in which the valve actuators are pressurized. In view of the limited amount of hydraulic fluid required, the most straightforward method might be to carry a precharged canister with sufficient fluids. Pressurized gas or chemical reaction gas generators could also be used to displace conventional fluids but the weight of the pressure vessel required is disadvantageous. If seawater was used as the control fluid a pump could be included in the LRROV. One of the advantages of seawater hydraulics is that, with the use of a pressurized gas system or chemical reaction gas generator, the LRROV would be lighter, easier to handle on the surface, and being more compact, more easily deployed on multiwell installations.

Initial considerations lead to a preference for an electric pump system being adopted for recharging the actuators. This will simplify turn round requirements and avoid the safety implications associated with pressurized gas systems or chemical reaction gas generators.

Telemetry

Telemetry requirements between the subsea tree and surface facility are minimal. The system selected would utilize the flowline for a communication path. The system is straightforward with acoustic energy being input into one end of the pipeline and a sensitive receiving transducer at the other. The system has been demonstrated over modest distances and adaptations to offsets in excess of 10 km would appear feasible without further technological developments although substantial detailing may be necessary. If the offsets were substantially greater, the range of the system could be extended indefinitely with the use of repeater stations, similar in concept to those used on transatlantic cables, on the flowline. These devices would be battery powered with a life of three to five years after which they would be replaced by ROV.

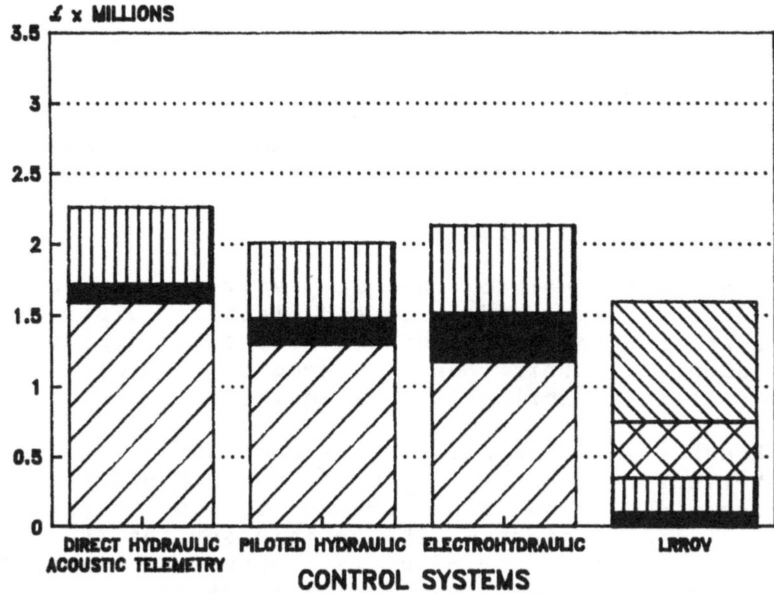

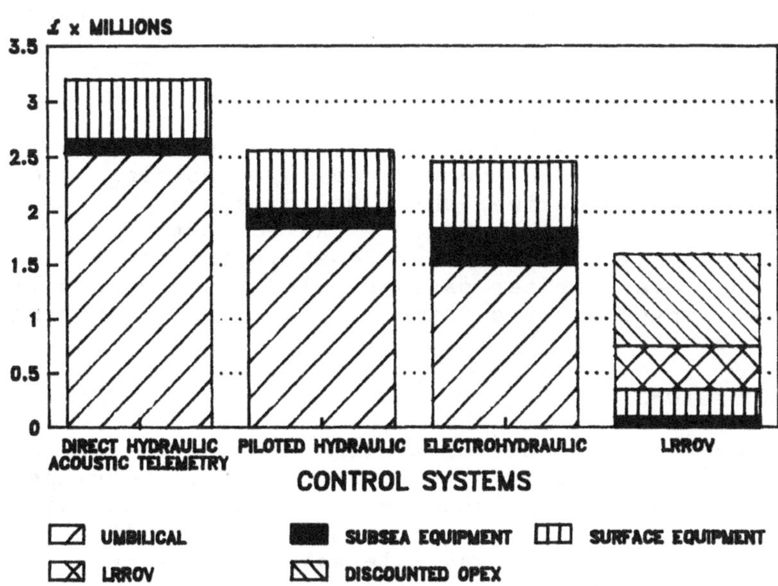

FIGURE 3. COST COMPARISONS

COST ADVANTAGES.

Figure 3 illustrates cost of alternative control systems for single well at 5 km and 10 km offsets. The direct hydraulic system includes acoustic telemetry for wellhead pressure; telemetry would be hardwired in the piloted hydraulic and electrohydraulic systems. Offsets less than 5 km are not considered since advances in directional drilling negate much of the incentive for developing controls to compete with existing technology at these shorter offsets.

Subsea costs include the dump valves, control module, acoustic package, hydraulic and electric distribution etc appropriate to the various concepts. Surface equipment includes electric and hydraulic skids, umbilical terminations, cabling etc as required.

Although all the systems have some associated operating (intervention) costs, including such things as umbilical inspection, these are assumed to be negligible for conventional systems. The vehicle serving the autonomous well requires periodic maintenance including minor refurbishment after use. An incremental operating cost of £100,000 per year for twenty years, has been discounted (at 10%) to yield a present value operating cost. Although it is true that the life cycle costs of the autonomous system need not include any umbilical abandonment costs, the effect of discounting makes this advantage negligible.

Umbilical costs dominate for the conventional systems. These costs could be reduced by combining installation with reeled flowlines which would reduce the advantage of the autonomous system. As offsets increase, the autonomous approach, whose cost is independent of offset, yields ever greater savings. If multiple satellite wells were served by the same LRROV, further savings would accrue.

CONCLUSIONS

A remote well control system, based on derivatives of existing technology, can be readily developed. The system is simple, operates in conjunction with largely conventional subsea equipment, and is based on adaptation of technology with an extensive underwater history. An autonomous system based on a long range remotely operated vehicle provides a lower cost control option for satellite water injection wells.

LOW COST SUBSEA SYSTEMS
FLEXIBLE PIPE - THE WAY FORWARD

E. MCGENNIS
Coflexip (UK) Ltd.
Commonwealth House
2 Chalk Hill Road
London W6 8DN
England

ABSTRACT Since the development of flexible pipe in the early 1970's over 2,500 km of product has been manufactured and then installed subsea. This has helped overcome many technical and commercial problems where conventional techniques were either unsuitable, unproven or not cost effective. Compared to regular linepipe the complex arrangement of plastic and steel is a premium product. Therefore it is important to have an understanding of flexible pipe performance to produce cost effective solutions to production problems. This paper looks at the principal areas where flexible pipe has been found to be the best solution and therefore producing the lowest cost solution. It also looks at future developments that should improve performance and reduce cost.

INTRODUCTION

Subsea systems can be configured in many different ways to optimise the economics of each reservoir. Within the range of options available, flexible pipe is able to provide a cost effective solution to a number of problems that can be encountered. These may be aggressive production conditions, complex subsea arrangements or small one-off systems. As flexible pipe is manufactured and tested on shore, the installation is simplified as there are only flange connections to be made between each length. As the cost of flexible pipe includes assembly, testing and certification of the finished product on reels, it is difficult to compare actual material cost per kilo or meter between flexible and conventional pipe. For this reason the best comparison is the actual installed cost considering also operation and maintenance.

SUBSEA SYSTEMS

A typical subsea system is made up of a number of key components. They include the various production and injection wells. These wells are then attached to flowlines which transport the fluids to the production system. Then there is the associated control equipment to monitor and operate the system. This paper focuses on the flowlines used subsea and how they can effect the economics of a project. The flowlines considered are the pipes used to transport produced and injected fluids from and into the actual well heads. With these various components, the overal cost of a

Volume 30: Subsea International '93, 13–22.

system is made up of the initial capital expenditure, the operating cost and the expected life. The capital cost covers the first phase of construction which includes wells, flowlines and associated equipment. This cost is traded off against the operation cost and expected life so as to achieve an economic solution. This economic solution will then be determined by the effective costs including interest payments when compared to the performance of the system. The performance will be the production achieved during the life of the system and the residual value. With most platform based systems the residual value is negative as the abandonment costs are high when compared to any resale or scrap value. With subsea systems the abandonment cost is lower and some equipment can be re-used. This may enable a subsea project to show a positive residual value. This is especially true of flexible pipe where over 700 km, have been recovered and re-used to date.

Competitive Solutions

Focusing on the flowlines for subsea systems there are many options to be considered and each one has a part to play as no one solution can meet the needs of every project. The range of options include conventional pipe lay where individual pipe lengths are welded and laid off-shore. Reeled pipelines, are an option used where the pipe is prepared onshore and then loaded onto a large reel to be laid in one continuous length. The use of bundles also involves on-shore preparation where the pipes are pulled into a carrier and then the entire unit is towed out to site. Flexible pipe is another option where the fabrication is done on-shore. The finished pipe is then delivered to site in long lengths which are flanged together as they are laid to the target. The choice of which system to use should always be based on the resultant economics of the project. As each method can bring particular solutions to certain applications it is useful to focus on the parameters that can make one option better than others. The parameters that determine the best option are largely length, diameter, layout and location. The longer the length, the more difficult it is to transport the entire finished product to site so this favours the conventional lay. Indeed most of the lengths laid in this manner are large diameter trunk lines rather than flowlines. As diameter increases the amount of pipe than can be stored on a reel becomes limited. If the field layout is complex it can be difficult to initiate and abandon flowlines in a small space. If the location has environmental problems such as sand waves or currents this can pose a problem of laying stresses and stability. On top of these considerations are the material options to meet the actual production fluids.

Flowline materials

The most common material in use is carbon steel. This is a proven system that offers low cost in the right conditions. If the conditions are severe however, the benefit of the initial low cost will be affected by increased maintenance costs for corrosion inhibition and repair as well as a shorter operating life. When pressures and temperatures increase and the level of sour components rises there is a marked decrease in the ability of carbon steel to resist corrosion. Even with high levels of chemical injection the effectiveness of corrosion inhibition can be reduced by turbulence, slug flow and high velocities. To overcome this problem it is common to look towards the various corrosion resistant alloys and materials. The options can include solid stainless steels, stainless steel cladding on carbon steel or flexible pipe. Other factors that affect the choice of the flowline system are the need for thermal insulation and heat tracing. To avoid wax and hydrate formation in flowlines it is sometimes necessary to retain as much heat as possible in the producing fluid. This is to keep the arrival temperature above the point where wax and hydrate can form. In certain cases it is not physically possible to reduce the heat loss to an acceptable level. This calls for either chemical injection or heat tracing. Heat tracing involves the input of heat along the pipe through electrical conductors wrapped around the pipe.

FLEXIBLE PIPE

Flexible pipe has advantages over other solutions for various subsea systems. The pipe is resistant to corrosion, easy to lay and has natural heat insulation. Looking at the basic design of flexible pipe for production systems there is first a stainless steel inner carcass which protects the pressure and fluid containing plastic sheath. This plastic sheath is then reinforced with carbon steel layers and the entire assembly is covered with an external plastic protective layer. As the plastic layers seal off the carbon steel the risk of severe corrosion is eliminated. For this reason the principal of using a plastic layer to contain the fluids eliminates the need for corrosion resistant alloys through the pipe.

Thermal Expansion

As the metallic components within a flexible pipe are wound in a helical manner the layers are able to slide and move relative to each other. This movement gives the pipe its flexibility but it also enhances the thermal properties of the pipe in terms of expansion. As the pipe gets hot the metal components expand as indeed do all metals. This expansion however is partially taken up by the radial expansion of the spiral wraps and partially along the pipe. This significantly reduces the thermal expansion of the pipe itself and therefore virtually eliminates the possibility of upheaval buckling under high temperature. The risk of buckling within flexible pipe is also reduced by the low bending stiffness. As the flexible pipe construction allows the layers to slide the pipe can bend easily and so it cannot form the long free spans necessary to cause upheaval. As other studies have shown, the amplitude of a buckle is proportional to the wavelength or length of influence. As this in turn is dependent on bending stiffness the risk of upheaval with flexible pipe is less than that of any other flowline system.

Termal insulation

As flexible pipe contains thermo-plastic layers to contain the fluids and protect the pipe structure, it has a natural heat insulation. Another factor that assists the thermal insulation are the various gaps that exist between the spiral layers. Therefore the flexible pipe structure can be considered as a set of metal layers sealed within two layers of plastic. In the high temperature systems where the inner sheath would be in Coflon the overall heat transfer of a standard eight inch flexible flowline would be around 15 w/m2 K based on the internal diameter. This is equivalent to a conventional steel pipe with a coating of 20 mm of neoprene. For production systems where only low levels of thermal insulation are required then flexible pipe can provide the solution with no additional cost.

Pipe laying

Considering the above factors, flexible pipe can be a cost effective solution to the problems of corrosion but it is also useful in the case of complex layouts or where there is only a single line to be installed. With conventional pipe much of the installation cost is on mobilisation, starting up the pipe lay and then abandonment. With flexible pipe the running cost of installation per meter is high but as start up and lay down are simple the overall cost can be reduced. This is true where there are many lines to lay over a short interval or where the lines need to be laid in a particular shape. Although the simplest way to lay out a field will be to have straight flowlines this is sometimes not possible as location of seabed objects can require the route to deviate. These can be items such as third party pipelines and cables, as well as natural obstructions. All these objects can require the flowlines to be laid in a curved path. As flexible pipe has recently been laid through the moonpool of a DP vessel this type of J. lay configuration allows control of the touchdown point to be very accurate and combined with the low bending stiffness of the pipe, tight curves can be laid along very narrow corridors. Again with the J. lay technique it is possible to lay flexible pipe directly beside one another. This is done by landing one pipe on top of the other and then letting it slide over to one side. Where there is a single line to be laid for a small project then the low mobilisation cost for a flexible pipe lay vessel can be an advantage. This is especially true if the single line is for water injection.

Water injection

As a flexible pipe for water injection is a simple structure of polyethylene and carbon steel the cost compaires favourably with carbon steel line pipe but with the added protection of the polyethylene layers which avoids corrosion. Although the actual cost per kilo of carbon steel is a lot less than flexible pipe it is the overall cost that determines the best solution. As flexible pipe can be laid off smaller vessels some which may already be mobilised for work on site the final cost can be competitive if the diameter and length are within a certain range. Considering conventional linepipe the work and cost to survey, weld, NDT, lay, test and certify the first kilo is very high. On a comparative installed basis the start-up cost per kilo for flexible pipe is less as long as the diameter and length are not excessive. As a rule of thumb flexible pipe can be a cost effective option to carbon steel where the total weigh of pipe is less than 500 tons. Although this is a useful guide it is important to note that the overall installed cost of flexible pipe is mostly material whereas rigid pipe is mostly installation. This makes the comparison very dependent of vessel availability and rates. Although flexible pipe can be competitive against carbon steel in certain cases the most widespread use is found in sour service environments. This is because flexible pipe can offer the corrosion resistance of exotic alloys at lower cost.

AGGRESSIVE SYSTEMS

The need to maintain production and utilise existing facilities has lead companies to look at ways of developing the more difficult reservoirs that were not considered in the past. As technology develops it is now possible to consider production from more aggressive environments with higher pressures and increasing levels of sour components. During the life of a field it is important that the production system can withstand these conditions and maintain an acceptable margin of safety during their life. Currently it is not yet economic to site a small production system over each well that can treat the produced fluid. For this reason subsea production flowlines are required to transmit the raw well bore fluids over various distances to the production plant being used. As the subsea flowlines are seeing the untreated well production they must be able to resist an ever changing combination of fluids, pressure and temperature. As the various electro chemical and material reactions are complex it is often necessary to estimate and extrapolate the behaviour of various pipeline materials. The most common material system currently in use is carbon steel pipelines. Using inhibitor it is possible to counter the effects of corrosion at certain pressures and temperatures. The higher the pressure, temperature and concentration of sour components the more difficult it is to guarantee the performance of the carbon steel. Factors such as flow, erosion and weld effects, all add to the complex reactions. To overcome these problems it is necessary to look at the stainless steel CRA's and flexible pipe.

Cost comparison

From the start flexible pipe has a cost advantage. First the cost of raw material. As flexible pipe has most of its weigh in carbon steels, thanks to the corrosion protection of the plastic the cost per kilo remains less than that for solid Duplex, even when manufactured, pressure tested and certified, flexible pipe remains cheaper per kilo when compared to CRA line pipe. On top of this advantage flexible pipe has a lower installation cost reinforcing the cost benefits per kilo. This would suggest that flexible pipe is at all times a better option than CRA pipe but there are two factors that reduce the effectiveness of flexibles at very small diameters such as below two inch and larger diameters above twelve inches.

Weight Factor

For the smaller sizes of pipe the factor that effects flexible pipe is the weigh per meter. Although the cost of flexible pipe per kilo is less than Duplex the amount of material required is higher. This is due to the nature of the flexible pipe design where the size of the individual layers within the structure cannot be reduced beyond a certain limit. This is due to the fact that the various forces such as pressure, tension and collapse are independant and taken up by different layers within the pipe. Another factor that comes into play is the relation between nominal bore and actual bore. With flexible pipe all internal diameters are actual bore as the nature of the design is to fix the internal diameter and increase the OD to achieve the pressure and stress levels required. This is the inverse of line pipe where the OD is fixed and the ID is reduced to increase the wall thickness. These factors combine to increase the flexible pipe weigh at small diameters around two inches.

Volume Factor

Where large diameter pipe is required there will be a physical limit to the amount of finished product that can be wound onto a reel. As the current limit of strain in flexible pipe is 7.5 per cent this requires the hub diameter of the storage reel to be around thirteen times the outside diameter of the pipe. As the pipe gets bigger the hub of the reel gets bigger. This then reduces the volume of pipe per reel. Above an outside diameter of approximately thirteen inches the weight of pipe that can be carried on the most common flexible reel in the North Sea drops below 190 tons. As the weigh per reel drops while the weigh per meter of the pipe increases then the number of reels required for transport increases dramatically. More reels means longer installation, more fittings and higher overall cost per kilo. This additional handling cost per reel becomes significant above twelve inch and reduces the effectiveness of the flexible pipe solution.

FUTURE DEVELOPMENTS

The long term decline in the price of energy in real terms has put tremendous pressure on the oil and gas industry to compete in the global energy market. With continuous demand to reduce costs much research has gone into flexible pipe to improve efficiency. In the near term a number of projects on design tolerances, materials, structures and use of longer lengths will help reduce flowline costs.

Manufacture

As with all fabrication work there are various manufacturing tolerances within the design of flexible pipe that reduce the efficiency of the end product. In an ideal situation the size of each component would be exact and therefore no excess would exist within the design. For this reason much work has been done to improve on the precision of manufacture so as guarantee the finished item without adding extra cost in terms of excess material. As well as improving manufacturing tolerances it is also hoped that a better understanding of the pipe mechanics can help refine the design. One of the most basic design parameters is stress as the stress levels within a pipe determine the thickness and size of the reinforcing components. Any refinements in allowable stress will help reduce weight and cost.

Stresses

When calculating stress the considerations will include the applied loads from pressure, tension and other factors. These stresses can be increased by the geometery of components where complex shapes can act as stress raisers. Stress can also be increased by manufacturing tolerance and expected material loss through corrosion or wear. Once the stress level is known then it must fall below the allowable limit from a safe design point of view. From an optimisation point of view however the stress level should fall right on the allowable level. This allowable level is then important to maintain an appropriate safety margin without excess. This allowable stress level is currently 50 per cent of the specified minimum yield strength (SMYS) according to the API RP 17B. Any increase in this allowable level will reduce the apparent safety factor but also reduce cost. As the safety factor applied is intended to cover the tolerances within the pipe the real safety factor in terms of real stresses can be different from the nominal safety factor used in design. As an example, the nominal safety factor depends on the accuracy of the calculation and models used as well as an understanding of the yield curve of the material. Any improvements in the accuracy of this data must only improve our ability to optimise the structures.

Materials

Flexible pipe is also looking towards new materials to improve performance. New thermo-plastic materials may offer higher temperature resistance at lower cost. Improved steel materials with new heat treatments may reduce the fabrication time and raw material cost. For deeper water some of the steel components may soon be replaced with lighter materials. On top of this, improvements are being made in the processing of the raw material and to increase automation this will lead to higher speeds and better quality as control operations can be built into the system.

Design

With the materials are new pipe designs which are being developed for improved efficiency. The teta wire design is now under test before approval for general use. This structure combines the function of two of the steel layers into one and allows higher pressures at lower cost. The Teta wire design also allows us to use a balanced 55 degree construction. This is the winding angle that produces no expansion or contraction under pressure. Currently 500 meters is under test and it has so far shown to be a very stable pipe construction. Work is also being done to improve the design of the pressure containing sheath to enable it to operate in production applications without the need for the current internal carcass. This will reduce cost and weight. There are also tests underway on a new design of heat tracing for flexible pipe. The new system involves an additional armour layer around the pipe which gives a large cross-section for the current. This allows us to obtain high current at low voltages, and so it is possible to obtain a large heat input per meter. This can be up to 1000 W/m of pipe.

Installation

Besides work on the pipe design, developments have also been made with the installation of flexible pipe. The most recent success has been the vertical lay spread which now enables the pipe to be lowered through the moonpool of a vessel in a J. lay configuration. This has increased the speed of connection between pipe lengths as the work is now done vertically within the derick before lowering the pipe. It has also improved the accuracy of pipe lay as the touchdown point is close to the centre of the vessel. This improves lay speed without risk to the pipe.

Diverless Connection

The efficiency of installation can also be improved with diverless tie-in and connections. Although this technique has been used extensively in deep water it has not really been applied in the North Sea within the 300 metres range of saturation diving. This has largely been due to availability of equipment and the need to mobilise divers for other subsea intervention. With improvements in the capabilities of the various remote operated vehicle's (ROV) and changes to subsea equipment design there is scope for a move away from manual work. Following the recent aquisition of Perry Tritech by Coflexip new developments can be expected and these will bring special benefits to the smaller projects when the economics cannot support the mobilisation and operating costs of additional vessels.

Recoverability

Another boost to smaller projects must be the ability to recover and re-deploy flexible pipe. This has been done extensively in Brazil but has yet to be a deliberate consideration in the economics of North Sea tie back projects. With the need to develop smaller fields there may be a case to tie back to a facility for just two or three years and then recover the equipment for use elsewhere. This reduces the overall cost of the operations. In a similar manner it may also be economic to have a long length of flexible pipe assigned to a fixed platform and which can be used to tap even the smallest pockets of oil within a certain radius of the facility. With a flexible pipe available on site already tied in to a platform then the cost of obtaining production or even a long term well test must be reduced every time the equipement is used.

Longer lengths

To improve installation efficiency further the connection time between reels needs to be reduced. Although it is possible to automate the actual process the best solution appears to be the complete elimination of joints. This will require the manufacturing and installation process to handle long continuous lengths of product. As the crosswound armours will lock up within a flexible structure, it is not possible to twist the pipe more than one degree per meter along its length. This eliminates the use of fixed baskets so the only real option to store and lay is a powered carousel. Although carousels exist up to 1200 tons the optimum solution for the flexible market would appear to be loads up to 2000 tons. This would allow continuous lengths up to 20 km. Plans already exist for the next generation of flexible pipe lay vessels and these plans are based on three independent carousels of 2000 tons each. Using a large carousel also eliminates the volume constraint of conventional reels so it will allow flexible pipe to become more competitive in the larger diameters over 12 inch. It will also reduce the overall cost by eliminating the intermediate flanges and the time lost making connections.

CONCLUSION

In summary flexible pipe has a part to play where its special characteristics can match the needs of a project. Where carbon steel is acceptable then flexible pipe can offer solutions for the smaller projects and where actual layouts are difficult to install. For problems of upheaval buckling and thermal insulation flexible pipe can be a competitive option but the real strength has been shown to be in the area of corrosion resistance. Where conditions are hot and sour then flexibles present a cost effective option to the conventional corrosion resistant materials. This comparison is best made on an installed cost basis where the advantages of flexible will be apparent. With the on going research work the target for flexibles is to offer the performance of exotic CRA material at the cost of carbon steel.

NORTH SEA FLEXIBLE FLOWLINES

FIRST BUILD DATE	OPERATOR	FIELD	DIAMETER RANGE INCH	HIGHEST PRESSURE PSI	TOTAL LENGTH KM
1976	Shell	Brent	4	10000	4.4
1978	Chevron	Ninian	3	5000	3.6
1979	Mobil	Beryl	6	5000	12.0
1979	Shell	Cormorant	3	5000	12.5
1981	Hamilton	Argyll	4	3000	17.0
1981	Texaco	Tartan	4	3000	3.1
1983	Hamilton	Duncan	2-8	5200	40.5
1984	Phillips	Ekofisk	8	4000	2.0
1984	BP (Britoil)	Beatrice	8	5200	4.9
1984	Hamilton	Innes	4-6	5000	17.1
1985	Oxy	Claymore	6	2000	3.6
1985	Oxy	Scapa	3-8	3500	8.5
1986	Statoil	Gullfaks	6	4000	13.7
1987	Amerada	Ivanhoe	4-8	3000	14.4
1987	Oxy	Chanter	6	3000	11.3
1987	Total	North Alwyn	6	5000	4.0
1988	BP	Buchan	2	5000	3.8
1988	Hamilton	Crawford	8	3000	2.6
1989	NAM	Ameland	7-10	4700	20.7
1989	Norsk-Hydro	COD 7/11	2-5	9800	13.5
1990	Norsk-Hydro	Gamma North	3-8	4400	25.0
1990	Saga	Snorre	3-8	5000	34.0
1991	BP	Magnus	6	4400	10.2
1991	Agip	Toni	4-8	10000	34.0
1991	Statoil	Statfjord	2.5-12	5000	14.0
1991	Shell	Draugen	3-10	3000	50.0
1991	Mobil	Ness	6	5000	7.6
1992	Elf	Claymore	6	3000	2.2
1992	Mobil	Camelot	3-6	3000	2.4
1992	Shell	Nelson	6-10	3300	30.0
1992	Kerr McGee	Gryphon	3-10	2900	16.8
1992	Norsk-Hydro	Troll	9-15	2500	42.5

"The Towed Pipeline Technique as a Means of Installing a Connected Structure."

C. J. Smith

General Manager - Pipelines

Kvaerner H&G Offshore Ltd.

(Inc. R. J. Brown and Associates (UK))

Introduction

Connected Structure refers to the ability of a submerged pipeline or bundle to be connected to a seabed installations then through to a surface vessel or loading facility. With the advent of exploration in progressively deeper waters the need for automatic systems or at best minimum diver intervention on the seabed is becoming a significant requirement. This necessity was recognised by R. J. Brown and Associates a number of years ago and design developments initiated at an early stage. RJBA has a long history of bringing innovative and cost effective pipeline installation methods to the offshore field development arena. Both for oil and gas installations in all parts of the energy producing areas of the world. This type of connectable system although of significant application for installation of single pipelines has particular use for the automatic installation of pipeline bundles contained within a carrier pipe.

Such a bundle would consist of production flowlines typically 6"dia. with a test line 6"dia. methanol/chemical injection lines 2"/1"dia., control/instrumentation umbilical and possibly heat

Volume 30: Subsea International '93, 23–36.

tracing. In addition the internal production flowlines may be insulated with solid cellular foam or gel. A carrier pipe of typically 24" /36" would contain the internal lines and probably the umbilical. The complete bundle system is therefore complex and for diverless connection there will also be automatically operating collet type stab-in connectors. RJBA's experience in the design, specification, fabrication and construction/installation supervision of these pipeline bundle systems is extensive.

Although not in deep water, RJBA designed such a system for automatic and diverless installation under the ice cap in the Canadian Arctic in 1976. This installation required all the consideration and innovative engineering that could be applied with the level of subsea equipment capability at that time. The bundle design and installation requirements were in fact ahead of the available technology and several areas of the installation requirements were prototype systems/instrumentation and installation equipment. From this significant early project to install a fully integrated connectable system in a diverless mode RJBA has been at the forefront of this type of installation with more advanced and sophisticated systems developed for installation in the Gulf of Mexico.

The connectable system is a significant part of the overall pipeline/bundle installation and its requirements have particular emphasis on the installation design/method. The pipeline installation method usually specified by RJBA for installation of the complete system is the beach make-up and bottom-tow method. It provides an effective method of pipeline/bundle fabrication without costly offshore work. Also a considerably reduced offshore installation period without the need for heavy column stabilised pipeline installation vessels with extensive saturation diving suites. This engineering approach is also followed to provide the maximum cost savings possible for the installation of complete subsea production systems. RJBA has designed and supervised fabrication and installation of a large number of bottom-tow installations which has generated considerable knowledge of applicable costs for materials, fabrication and installation. This data base being used to refine further cost effective bottom tow and the associated connectable systems. Once the pipeline bundles are connected however a method of transferring product from the seabed to the surface reception/holding vessel is required. Through development work by RJBA and on a recent project for Chevron Research a Self Erecting riser has been proposed. This riser being an integral part of the towed system with deployment after interface of the towed pipeline/bundle system to the seabed template or manifold. The system is of particular significance for the transfer of satellite wellhead product to a floating production Tethered-Leg or anchored surface production platform.

By the nature of this design approach the design and construction supervision work is specific to the pipeline/bundles needs. Although a reliance on established equipment and vessels and operational procedures is maintained certain aspects of each installation require development and prototype aspects. The risk of such novel design and installation is however mitigated and controlled by the close cooperation between RJBA and the Client/Operator organisation. Due to the innovative approach to this type of pipeline bundle design and installation the concept of Alliance Participation between Client/Operator and Design/Construction Contractor has not been applied to date. However towed pipeline installation projects have been carried out wherein complete responsibility for the design and installation against penalty and bonus for milestone achievement was applied.

The method of installation by bottom-tow has been described in detail in previous papers presented by R. J. Brown and Associates. The intention therefore is to discuss further particular aspects of the bottom tow technique particularly in the context of connecting a complete towed bundle into a seabed structure and the method of transferring product to the surface vessels

Towed Connectable Systems

Various techniques are employed by the pipeline bundle fabricators and installers to connect the pipeline bundle to the seabed structures. The method developed by RJBA is the deflect to connect method. Connection of the pipeline bundle is achieved with minimum residual load on the connector head by adopting the deflect to connect procedure. The pipeline bundle is towed into a target area on the seabed then buoyed of the seabed by releasing chains along approximately 300mtrs of the leading end. By connecting a pulling wire to the leading end of the pipeline bundle and a hold back wire the bundle is deflected through approximately 90 degrees into alignment with the receiver on the base of the seabed installation. Connection between the pipeline bundle and seabed installation being made by a towing head designed to interface with a compatible face plate on the seabed installation. Once the pipeline bundle has been deflected and pulled into the receiver a stabbing pin locator is locked in position and a collet connector or movable mating face is activated to seal the bundle flowlines, chemical injection and instrument/controls umbilical together. Locking of the stabbing pin locator removes any residual stress in the bundle to ensure there is no strain on the flowlines/umbilical connections. The buoyant end of the pipeline bundle is then stabilised on the seabed with mattresses or localised gravel dump.

Experience has now been gained on a number of towed pipeline installation systems. The particular advantage of these methods is in the ability to fabricate and pretest the complete pipeline bundle onland and not be subject to the offshore environmental conditions. The fabrication and construction plant such as heavy lift cranes are only utilised over specific days and not on long contract periods. This is in marked contrast to the complete construction spreads based around 3rd Generation Laybarges and MSV construction support vessels incorporating extensive saturation diving facilities. Such vessels also have extensive mobilisation and demobilisation periods and although they have a high sea state work capability due to their column stabilised barge structure they are still subject to weather and mechanical downtime.

The basic principal is the same in each of the towing method variations. The pipeline assembly is towed by a high power (bollard pull) ocean going tug or large offshore supply vessel with a smaller tug maintaining a holdback tension on the towed string. The variation in methods concerns the type of towing whether surface, subsurface, mid-depth, off-bottom or on-bottom each method having its advantages and failings. These methods fall into two distinct groups either tow through the water mass or tow on the seabed. Particular advantage of the former group being that a seabed survey is not required for the pipeline towing route where this is a necessary requirement for the on-bottom tow method. The cost of this seabed route survey however must be equated against the cost of buoyancy or additional carrier pipe diameter required for the subsurface/mid-depth tow methods. By restricting the size of carrier pipe a specified flowline bundle seabed weight is determined which can be selected to permit towing of the complete bundle by the Clients regular on-site supply/anchor handling vessels.

The towed method of pipeline bundle installation has been developed and promoted by R. J. Brown and Associates and has a number of advantages over the subsurface/mid-depth tow methods. In two recent pipeline bundle installation projects the fabrication and assembly works was arranged for maximum flexibility and cost effectiveness. With the ability to respond to other field associated delays and keep the cost impact from consequent additional labour and equipment time requirements and cost to a minimum. The complete pipeline bundles were fabricated and assembled on a beach head and pulled off the beach into the water then towed along the surf zone area into deeper water. A particular aspect of these installations was that they were to be installed in deep water in excess of 350mtrs. and in a predominately diverless mode. The final connection being achieved by the deflect-to-connect method, as described earlier. By utilising the facility of a long straight uninterrupted beach head at Matagorda Peninsular, Texas, for this fabrication assembly and testing/precommissioning work minimum impact is caused to the environment. This is in marked contrast to the purpose built make-up and assembly facilities constructed laterally to the beach head with all the associated, buildings, storage yards and support facilities. Two such pipeline/bundle make-up yards exist in the UK at Wick and Tain, both in Scotland, operated by separate construction contractors. Both of these contractors fabricate towed bundles for the subsurface and mid-depth tow methods the former being operated by Rockwater and the later by Costain (Land and Marine Engineering).

Suitable sites for fabrication of pipeline bundles along a beach head in the UK have been identified. Fabrication of bundle lengths to 11km in one length could be achieved on the South East coast of England. Several sites were evaluated for a major North Sea Operator for tow out and installation of a large carrier pipe/bundle assembly. These sites are within easy reach of comprehensive engineering facilities and site engineering personnel, support services, ie. transportation, plant/equipment hire, accommodation, subsistence, etc.. Further beach sites have also been identified along the East Coast of Scotland a particularly useful length of linear beach head being adjacent to Montrose, in excess of 11Km.

The towing requirements of each method differ considerably and have a direct effect on the final installation and positioning on the seabed. For each of the subsurface/mid-depth tow methods the buoyancy and hydrodynamic towing characteristics are significant aspects of the towing and installation procedure. The surface tow method requires the pipeline bundle to be supported by surface buoyancy in the form of variable buoyancy bags or collapsable spheres. Mid-depth tow requires that the towed system acquires a neutral buoyancy during towing and that the depth of tow can be controlled for the optimum towing efficiency. Also the maximum flexibility in positioning on the seabed at the installation location.

Integral Self Elevating Risers

From work conducted by R. J. Brown and Associates on the Placid Oil Company, Green Canyon Block 29 and Enserch Mississippi Canyon 441 deep water pipeline bundle installations in the Gulf of Mexico an innovative connectable system is under development.The system consists of the seabed to surface riser being fabricated and supported by the pipeline bundle assembly as an integral part of the towed system.

Once the carrier pipe/bundle has been fabricated, the leading end of the assembly would have a continuous buoyancy tank, approximately 300mtrs in length attached on top of the carrier pipe. This buoyancy tank is also the riser and contains the flowlines, chemical injection, instrument/control umbilical for interface to the surface storage and tanker transfer facilities. At regular intervals along the buoyancy tank/riser would be lengths of chain terminated onto the tank but folded and retained together with wire or overcentre release. When the complete assembly is towed offshore and to a position adjacent to the seabed template/manifold the chain ends are released, as described earlier, causing the tank buoyancy to balance the suspended weight of the chains. This then allows the assembly to be deflected laterally with a low force and hence permit a final minimum stress connection to the receiver. The lead and trailing end tow head structures will be buoyed off-bottom, slightly positively buoyant, by the attachment of buoyancy tanks to reduce the total tow force required by the towing vessel. The buoyancy of the overall system will be design to be trimmed to permit towing of the complete assembly with a bollard-pull of approximately 150 tonnes. The typical on-bottom submerged weight of the system would be approximately 9Kg/m to provide lateral stability against transverse seabed currents during the towing operation.

With the pipeline bundle and buoyancy tank/riser located and connected into the seabed template/manifold the riser would be released to the vertical location. Various assessments and model simulations have shown that the riser would rise at a controllable rate and not impose excessive bending or displacement frequency on the deployed system. A procedure of sequential release of the riser would be determined by scale trials during the detail design phase. The motion characteristics of the riser during the uprighting operation will have limiting aspects with regard to the prevailing environmental conditions. Lateral currents due to surface waves and wind in addition to the movement of currents through the water column will be required to be collated and evaluated. The method of connection and transfer of product from the seabed bundle to the riser has already been considered in detail and various swivel and product transfer systems specified. Such swivel systems as ball and socket fabrications of exotic materials to simple three chain tripod arrangements have been proposed. It is generally considered that the simpler the system the better with short jumper hoses between the seabed bundle and riser internal cores. To maintain tension on the swivel a system of installing or pulling a buoyancy tank down the length of the riser from the surface will generate sufficient longitudinal and lateral resistance to excessive oscillating movement about the swivel joint. This will also resist bending of the riser and the fatigue effects due to vortex shedding along the risers vertical length through the water column.

Recent Self Erecting Riser Projects.

The concept of an integral riser as part of a towed pipeline/bundle system is new it has been proposed by the established designers and fabricators of towed systems. In certain applications as part of an integral drilling towhead and drilling template or manifold system interfacing remote satellite wellheads. The riser is then released from its position along the pipeline/bundle and raised to the surface and receiving vessel or loading facility.

Recent projects conducted by RJBA show a development from concept to feasible design arrangement in accordance with the offshore Operators requirements. A number of Figures 1 to 5 are shown at the end of this paper and indicate the progression of the design from 1986 to 1992.

The Placid Oil, Green Canyon development, Figure 1., required a semi submersible floating production system a 1,450ft. free standing riser a 24 well template, dual export pipelines and three flowlines. Placid Oil,s conclusion on the installation of the pipelines was to utilise the bottom-tow method for less field congestion during pipeline installation and diverless connection. This project showed that the concept of a Self Erecting riser would be feasible being deployed from the back of the towed pipeline/bundle assembly.

Further developments were conceived in projects shown in Figures 2. and 3. respectively Tatham Offshore, Mississippi Canyon Block 194 and Enserch Mississippi Canyon 441.

The recent projects for Chevron Research in the Gulf of Mexico GC205, Figures 4. and 5. have concerned respectively,'Deep Water Pipeline Installation and Riser Tie-Ins' and 'Deep Water Production Self Erecting and Drawdown Risers'. Each project has considered and developed the application of freely erecting risers as part of the initially towed pipeline/bundle system.
The objective of the latest project was to prepare sufficient

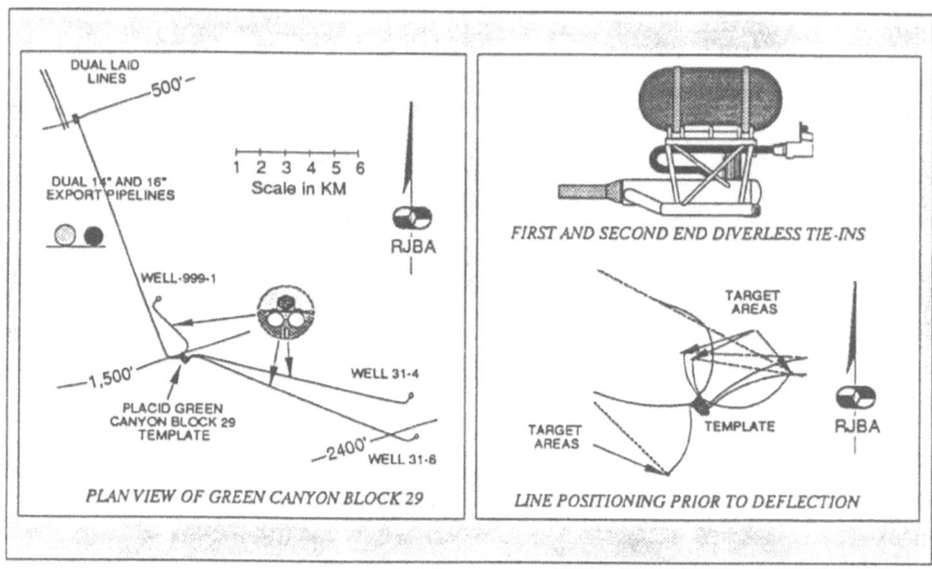

Figure 1

Deep Water Development Export and Flowlines

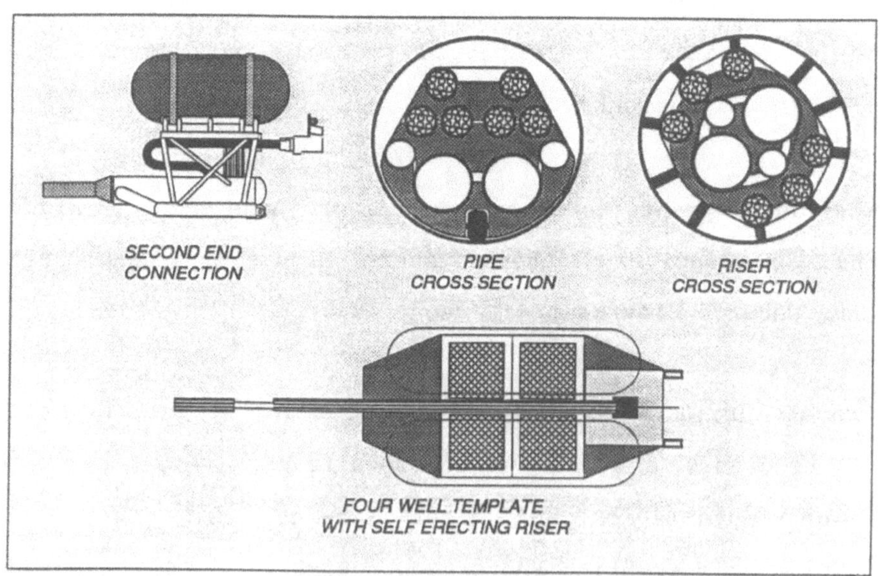

Figure 2

Mississippi Canyon Block 194 Pipelines

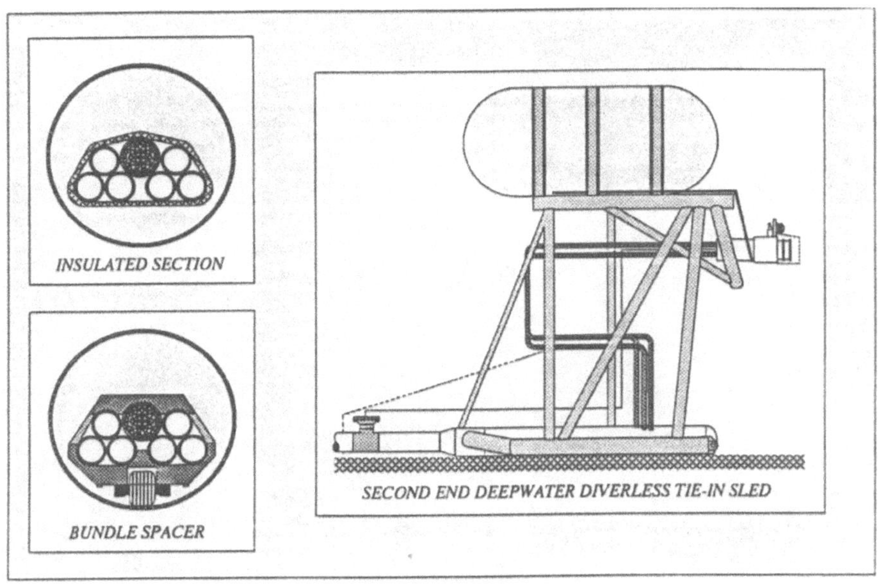

Figure 3
Deep Water Template Tie-ins

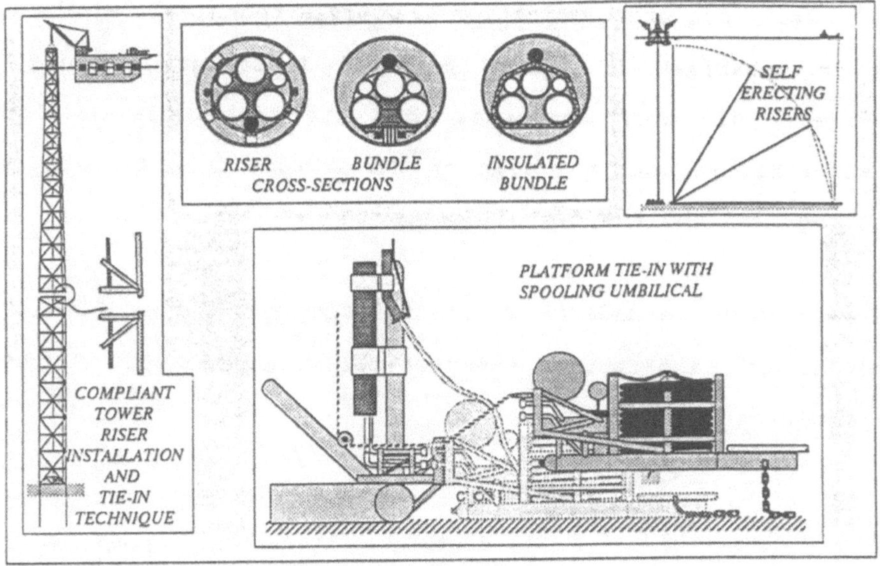

Figure 4
Deep Water Pipeline Installation and Riser Tie-ins

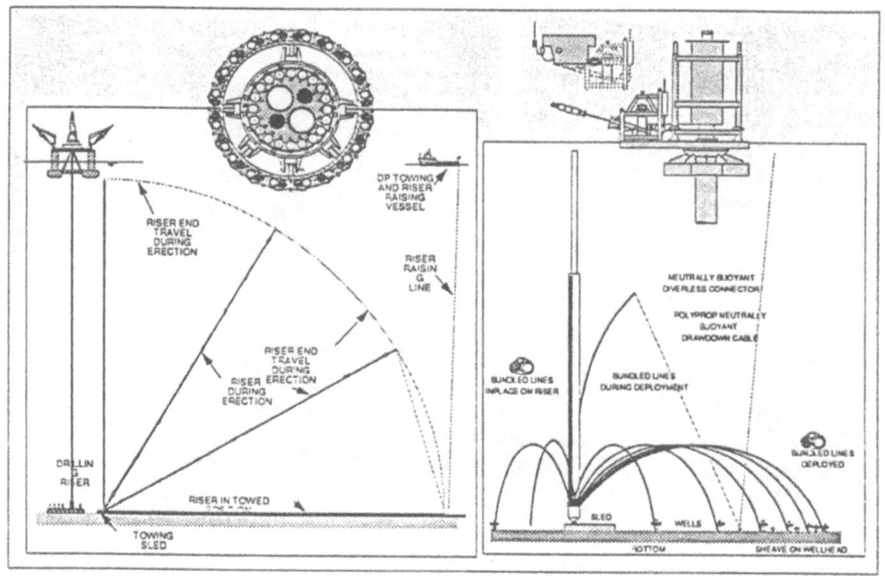

Figure 5

Deep Water Production Self Erecting and Drawdown Risers

design and procedures definition to confirm technical feasibility and to establish if lower costs of installation could be achieved. This work confirmed that the techniques developed considerably reduced the size of the surface spreads, offshore time required and the overall costs.

Acknowledgment is made to Placid Oil Company, Tatham Offshore, Enserch Corp. and Chevron Research for reference made to their projects.

Kvaerner H&G Offshore Ltd.

January 1993

FUTURE RISER REQUIREMENTS - A LOW COST SOLUTION

ALAN BRACKENRIDGE
Subsea Engineering Manager
OGL Engineering Division
Regent Centre
Regent Road
Aberdeen AB1 2NS

ABSTRACT

The development of optimal solutions to satisfy the riser requirements for new platforms has considerable financial implications during platform construction and platform operational life.

Techniques recently developed by OGL offer substantial cost, safety and technical benefits over traditional riser solutions and draws on practical experience gained from over ten years development and installation of retrofit riser systems.

INTRODUCTION

Platform risers are key components of all offshore production facilities having major economic and safety significance. The riser provides access from the sea bed to the platform facilities for:-

- Product import
- Export pipelines
- Platform utilities (control and power lines)
- Subsea satellite developments, (produced oil and gas, test lines, water injection and control umbilicals).

A riser is usually a pipe which connects between a subsea pipeline and the platform topside and can be made from rigid or flexible pipe in the size range from 2" DIA to 36" DIA depending on requirements.

Volume 30: Subsea International '93, 37–47.

The riser design accommodates substantial environmental loading, particularly in the splash zone. To accommodate the environmental load risers of smaller diameters require additional structural support which is provided by :-

- Additional clamps to span between main jacket members
- Located inside a carrier pipe called a caisson which acts as a structural member to provide protection

Risers are of key importance to platform safety as their failure can compromise the complete production facility. For this reason many safety considerations are incorporated into riser designs.

These include:

- Accidental Vessel Impact Protection
- Increased Requirements For Corrosion Protection
- Structural Strengthening
- Subsea Safety Valves
- Passive Fire Protection

Riser design is covered by numerous industry specifications and government legislation, included in which, are recent recommendations by the UK Department Of Energy which requires risers to be protected from accidental vessel impact damage.

A major requirement on all new production facilities is to accommodate current, future expected and future speculative risers.

Traditional methods for accommodating future riser requirements on production platforms have been:-

- do nothing, and completely retrofit at a later date if required
- fit best guess risers requirements during construction
- fit 'J' and/or 'I' tubes during construction allows later risers to be installed
- fit appurtenances onto the jacket structure to assist riser installation

All these methods have economic and technical disadvantages. OGL have delivered new techniques based on retrofitting riser systems onto existing jackets which have considerable technical and economic advantages over traditional methods.

The paper will review the traditional techniques for providing future riser access, both technically and economically. These techniques will be compared with the recent developments in platform riser technology which have recently been deployed on several new jackets.

HISTORICAL

There is no industry standard methodology for the accommodation of risers onto jackets. Each operator and jacket designer have developed methods based on ideas which were developed over 20 years ago. Future riser requirements are often given very little consideration during the initial jacket design leading to risers and future riser requirements being fitted as an afterthought, where the jacket configuration allows. This approach invariably leads to a compromise design resulting in non optimised riser systems and increased operator cost.

Traditionally additional and spare riser capacity is incorporated at the design stage of new production facilities by one or a combination of the following methods:

- **Install On The Jacket Known Riser Requirements**

If riser requirements are known during platform construction then these are normally installed.

Risers installed during platform construction have several concerns:-

- During the life of the riser if a problem develops or the riser is no longer required no remedial action can be taken.

- Smaller risers require clamping every few metres which results in an increase in platform load, weight, and construction costs.

Where a substantial number of risers are involved, additional structural members may be required. Figure 1A shows a cross section of risers and support steelwork.

This system within the jacket boundary has the smaller risers exposed to environmental and impact damage.

Alternatively smaller risers can be installed in a caisson. This method is considerably safer with the riser protected from environment and impact damage, but the principal concern is inspection of the risers.

If the environment in the caisson is controlled then the risers integrity should be assured. However if a problem does develop then one riser could comprise all the others.

Traditional caisson riser systems do not have the facility to retrieve and replace an individual riser. This will have the effect that a fault on one riser could result in a lengthy and costly shutdown of the complete system.

- **Second Guess Future Riser Requirements**

Using similar methods to those described above additional risers can be installed on the jacket for later tie-ins. Future riser requirements are guessed with no guarantee they will meet future needs. (materials, size, pressure). In addition if the riser is not brought into service for many years the riser condition may deteriorate if not fully maintained.

The weight and cost of installation of future riser systems may be considerable and permanently load the jacket even if not used.

- **Incorporate 'J' or 'I' Tubes Onto The Jacket**

'J' Tubes are essentially small caissons with a large 90° bend at the lower end which allows a riser to be pulled inside the 'J' tube. During the installation of rigid risers high stresses are developed which cause the riser material to yield as it passes around the 90° bend. This operation also imposes considerable stresses on the 'J' tube. The 'J' tubes are permanently installed on the jacket adding considerable weight and loading whether used or not. A 'J' tube has limited capacity and usually accommodates one riser per 'J' tube as riser bundles are technically difficult to install. Over time the condition and structural integrity of the 'J' tube can deteriorate if not fully maintained.

'I' tubes are similar but with a shorter bend which allow only flexible lines to be pulled through.

Typical cross section of 'J' or 'I' Tubes as shown on Figure 1B.

FIGURE 1A - SUPPORTED RISERS

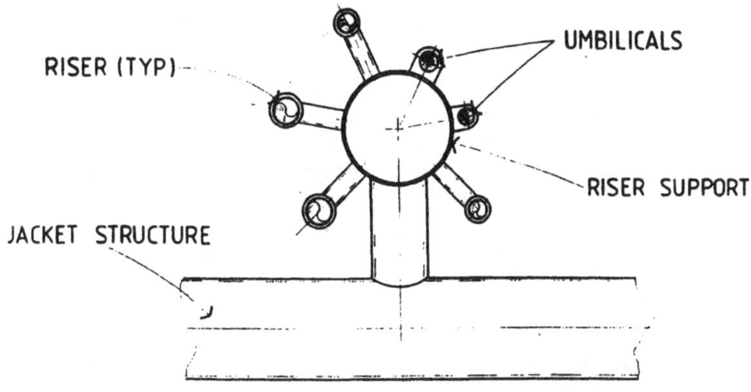

FIGURE 1B - RISER IN 'J' TUBES

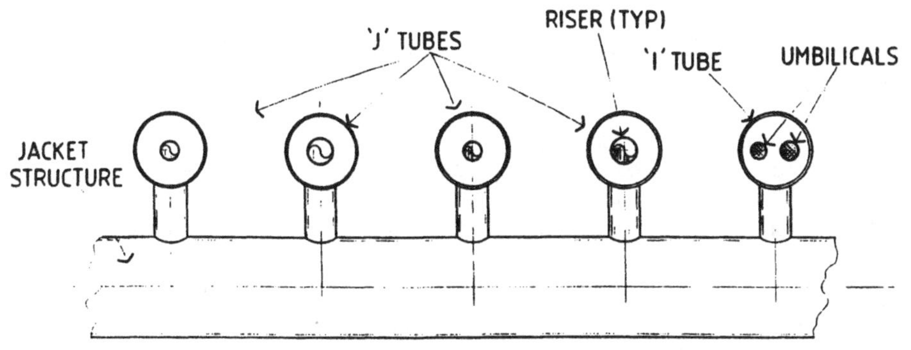

FIGURE 1C - RISERS IN RETROFITTED CAISSON

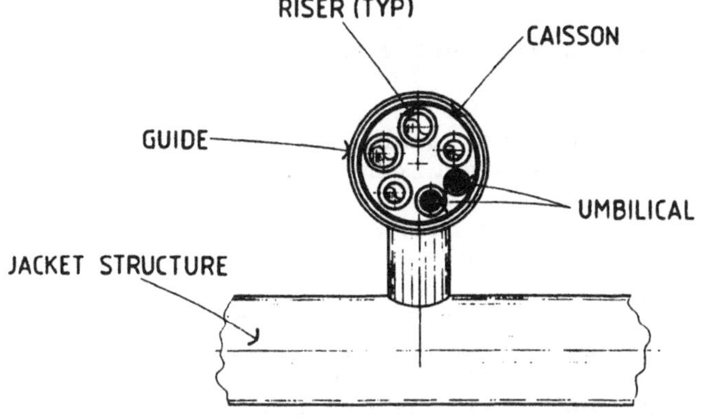

• Fit Appurtenances on the Jacket to Assist Riser Retrofitting

Typically bolted flange plates are installed during jacket construction which allow a riser to be installed later. This system is limited to retrofitting on the outside of the jacket and involves extensive diver and lift operations, at considerable cost. With the riser on the outside of the jacket it is subject to vessel impact damage and will require additional protection to meet current directives.

Typical layout of traditional systems is shown on Figure 2.

NEW METHODS

A new approach has been developed and is now in operation which optimises future riser requirements at minimal cost.

This is an integrated approach which assess riser requirements and develops a selection of options which are analysed to establish the best total solution both technically and economically.

To establish riser requirements they are identified in four categories:

- Required: size, rating and service - known
- Very probable: size, rating and service - fixed
- Possible: size, rating and service - best guess
- Spare: size, rating and service - unknown

The main elements of this approach include:-

- Think retrofit at the earliest possible design stage of the jacket and platform deck. An early start would permit access through deck to be vertical, with vertical caissons and a significant reduction in installation cost.

- Establish total riser requirement both known and future unknown.

- Establish the best combination of riser access methods to meet requirements.

- Design all risers below 12" inside the retrofit caisson. (Larger risers can be retrofitted without a caisson).

- The riser system design is installation driven and based on the substantial experience gained from previous installations of retrofit riser systems.

FIGURE 2 'TRADITIONAL' RISERS - NEW JACKETS

FIGURE 3 ALTERNATIVE RISER SYSTEM - NEW JACKETS

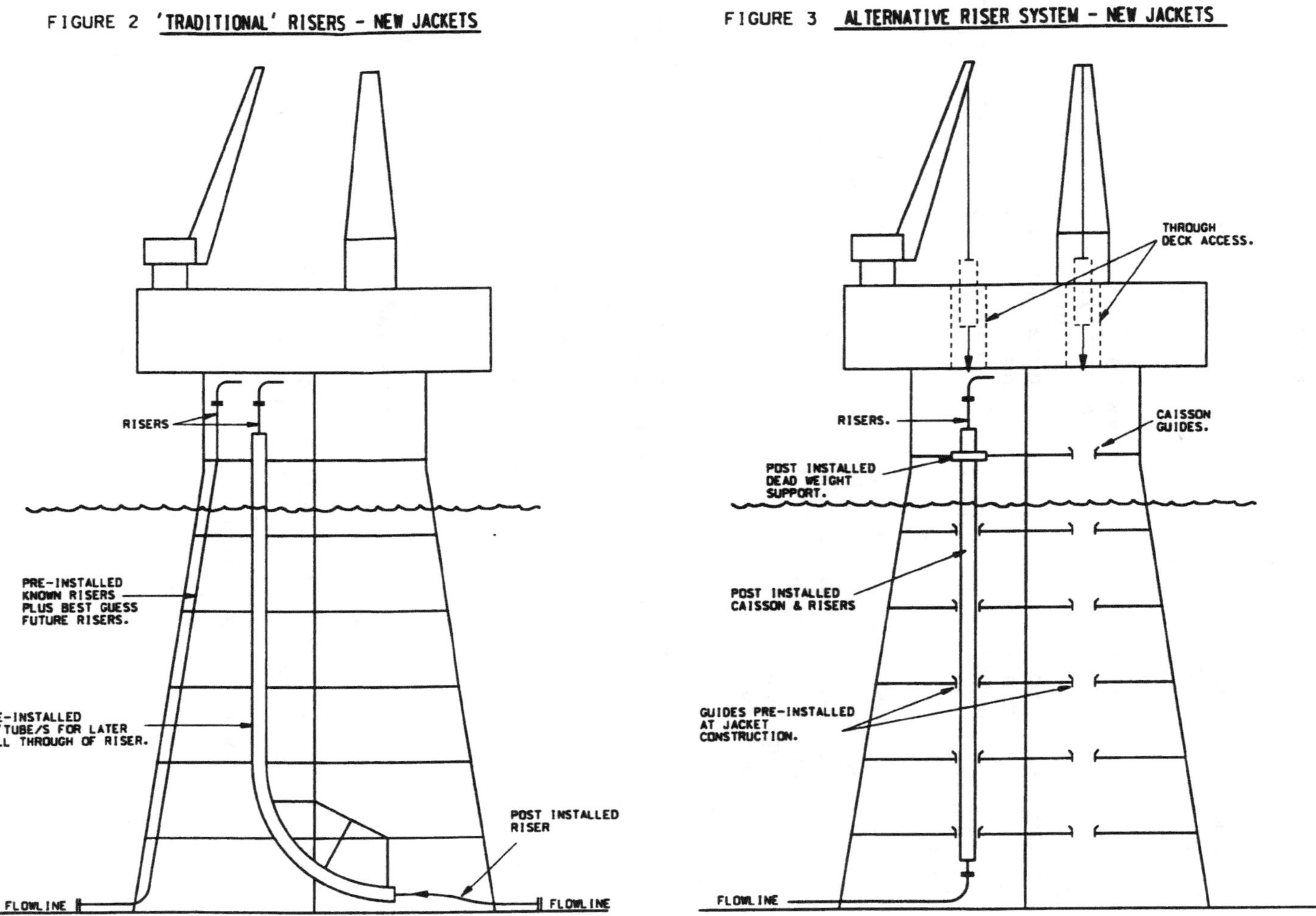

- Accommodate and reserve space/transparency in jacket strategic locations for retrofit.

This technique will utilise existing traditional riser access methods and the OGL system for retrofitting onto existing platforms, adopted for new jackets.

By incorporating some minor changes to jacket and topside layout, a complete retrofit riser system ie. typically eight risers can be installed. This system allows total flexibility of riser selection with the system installed when it is required with minimum modification and additional weight to the jacket at construction. Typical additional requirements to a jacket at construction will be caisson guides at principal jacket levels.

This riser system can accommodate all riser types and sizes from 2" DIA to 36" DIA with flexible and rigid risers or combinations, power, umbilical and other service risers. All water depths can be accommodated from platforms in shallow water to deep water where diverless flowline tie-in techniques have been developed.

In this system the smaller risers are housed in a caisson for environmental protection and support. The system is installed in sections using existing platform facilities. The risers are installed individually and are retrievable.

A typical caisson cross section is shown on Figure 1C.

A typical system layout is shown on Figure 3.

COST COMPARISON

It is difficult to make accurate cost comparisons for future riser systems, however a comparison is made on an assumed future riser requirement.

A cost comparison has been made to compare four different solutions for a future riser requirement:-

2 x 8" riser rigid
2 x 6" risers rigid
2 x 6" risers flexible

The riser is installed on a steel jacket at a water depth of 100m.

Four future riser scenarios are considered:-

1. Make no provision of future riser requirements and install a caisson riser system when required.

2. Make best guess at future riser requirements and install a caisson riser system during construction.

3. Make best guess at future riser requirements and install six 'J' tubes during platform construction for later pull-in of risers.

4. Make provision for OGL riser access method.

The engineering, construction and offshore installation costs were estimated for

- Cost during platform design
- Cost during platform construction
- Cost to complete the riser access when required during platform life

The results are shown on Table 1 herein:-

- Riser can be specified during platform design, they should be fitted during construction

- To make no provision for future risers results in high cost if retrofit risers are needed, approximately x 2 the cost of any other alternative

- Traditional approach of fitting 'J' tubes during construction results in increased riser costs and limited access and flexibility

- The OGL solution gives maximum flexibility at minimum cost for marginal increase in construction cost the complete retrofit riser system can be installed at a later date

TABLE 1

Activity		-1- No Work Installed as Req'd	-2- Install Riser During Construction	-3- Install J Tubes	-4- OGL System
Engineering	Up front		165K	125K	75K
	Complete	325K		195K	150K
Construction	Up front		410K	355K	105K
	Complete	730K		250K	360K
Offshore Installation	Up front				
	Complete	3150K	600K	1060K	740K
Cost Up Front		---	575K	480K	180K
Cost to Complete		4205K	600K	1505K	1250K
TOTAL		4205K	1175K	1985K	1430K

INSTALLATION OF THE RISER SYSTEM

• Installation of the Caisson

The caisson is installed, complete with the conduit bundle by lowering the caisson assembly in sections from the platform crane or drilling derrick through the platform and into guides in the jacket. The caisson sections are installed in lengths of (12m) which are connected together using a conductor connector or welded. The assembly is lowered by repeating this procedure until the caisson is installed between the seabed and dead weight anchor support below the topside. The installation is inside the platform.

• Rigid Risers

The risers are installed individually from the drilling derrick in pre welded lengths of 90ft. The lengths are welded together in the derrick and run down through the topside module and into the caisson. Each riser has a dedicated conduit which runs the whole length of the caisson and separates the risers.

- **Flexible Risers**

Flexible risers and umbilicals are installed using standard flexible line pull-in techniques either up or down the caissons.

SUMMARY

New and alternative techniques have been developed to optimise riser requirements for known and future riser in new jackets. These techniques are now being utilised and early consideration of riser requirements will result in the optimised system which increases flexibility at reduced cost on new jackets.

MANTA - THE BOTTOM LINE SUBSEA PRODUCTION CONTROL SYSTEM.

ROBIN BARNES.
Myrmidon Subsea Controls Ltd,
Broadbridge Industrial Park,
Horsham, West Sussex, RH12 3JR.

ABSTRACT.

This paper presents a lateral approach to subsea production control, the intelligent umbilical or **MANTA** concept. The concept eliminates the need for a discrete subsea control package and a topside hydraulic power unit by placing both in the subsea end termination of the control umbilical. The bottom line is that this approach provides superior system reliability and can be shown to have major economic advantages.

The object of the exercise is to provide a system which will control the production of hydrocarbons from a remote location on the sea bed cheaply, effectively and reliably.

The **MANTA** concept is based on a novel combination of proven technology. This effectively provides for a high reliability system combined with significant reduction in the cost of ownership. At the same time, the concept is, by its very nature, able to take full advantage of all the very best aspects of up to the minute technology.

The **MANTA** will be found particulary attractive for applications in marginal remote field developments.

INTRODUCTION.

The production of hydrocarbons from under the oceans of the world is now considered an everyday event. The MANTA concept is a logical step forward in the evolution of underwater production control. The MANTA approach has evolved from the integration of a combination of already proven subsystems. This amalgamation of ideas has provided a system which has both improved reliability and a clearly demonstrated reduction in the cost of ownership.

Volume 30: Subsea International '93, 49–65.

The MANTA is seen to meet a number of the major industry requirements for a sound but flexible basic subsea production control system package.

The subsea control systems being installed today are, in a number of ways, at least 10 years out of date. Technology has advanced, the industry meanwhile would seem to have stood still or in some cases gone backwards. To some extent this is understandable, Oil Companies are in the business of producing oil and gas - not hardware. Any field development is a risk. Limiting the initial investment helps reduce the risk to what would seem to be a sensible minimum.

The problem with this approach is that while the Company wants 10 year old fully qualified technology, the initial budgets are limited. Suppliers therefore in order to be competitive, are forced to look for ways of reducing costs and/ or of meeting specification which, again on limited budgets, have often been generated by very inexperienced personnel. The nett result of this conglomeration of sometimes conflicting ideas and requirements is generally a further compromise to the 'standard' equipments performance.

Current and traditional systems exhibit inherent weaknesses which are amplified by economic pressures. Platform space allocation is nearly always marginal. J tubes are almost always too small. The size and weight of both surface and subsea equipment invariably presents major installation and deployment problems and, subsea equipment failures, although not an every day event, are nearly always attributed to either infant mortality, tramp contamination in the hydraulic system or degradation failure of underwater installed interfaces.

The Manta concept provides a novel and lateral solution in overcoming most of the major limitations of today's subsea control systems, while at the same time presenting significant economic advantages to the end user.

This paper presents the major technical and economic benefits of the concept and provides comparisons with similar aspects of existing equipment.

It is a fact that 'Subsea' is still a relatively new industry and it should never be forgotten that the environment being dealt with is alien in nature. Care and attention to detail must therefore be exercised at all times.

Before describing the 'MANTA' concept in detail, it may be helpful to consider some of the more general aspects of subsea production control, its evolution over the last decade, and some of the more traditional methods of control system equipment installation.

EVOLUTION.

During the last ten years demand has increased sharply for suitable equipment to control and monitor subsea production. This demand has invariably been met by provision of equipment based on variations of existing 'proven' designs. This has of course included their inherent limitations and faults. The system has then been improved by incorporating 'nice to have' improvements which have in turn built-in additional faults.
The consequence of this approach is that the overall reliability of the system has suffered and, to some extent, the industry has stepped backwards.

In 1976 the first single satellite early production tree, using a hard wired electro/hydraulic control system was installed in the North Sea. This was the culmination of a very intense international development programme with some very good ideas and a somewhat limited budget. It is now a matter of record that the control system did in fact work for the expected design life. Failure of control, when it finally occurred, would seem to have been at a poorly conceived subsea primary interface.

In 1978 a prototype single satellite well, using inductive couplers and multiplex electro/hydraulic control was installed in the North Sea. For system reliability, the development used dual redundant hydraulic supplies and electrical power/signal lines. All umbilical service lines terminated at the christmas tree interface connectors. In order for the system to perform its function, the muscle power, in the form of high pressure hydraulic fluid had to pass through five underwater mateable interfaces. The electrical power and signal functions had in turn two such interfaces to traverse.

FIGURE 1

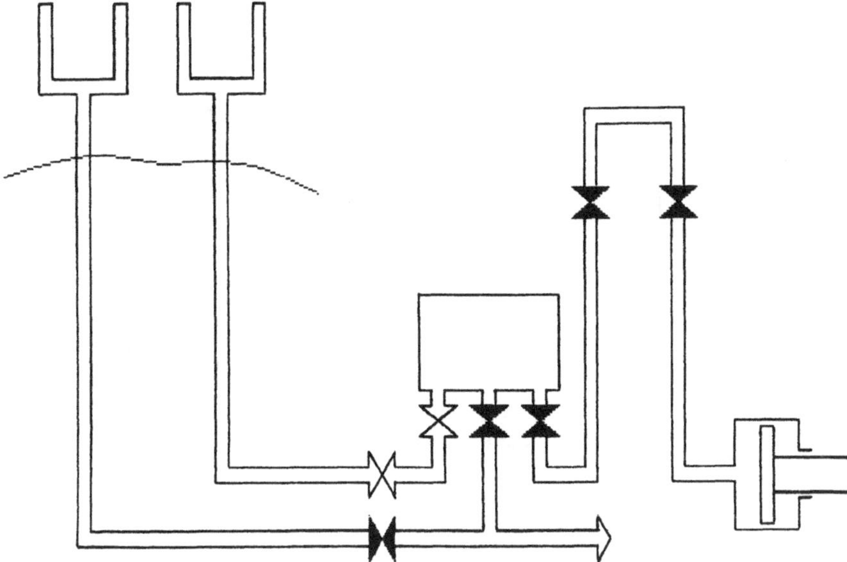

In 1982 a project involving the diver-less installation of a series of single satellite wells was undertaken. By this time some major advances in the performance and capabilities of remotely operated tools and vehicles had occurred. This, together with growing awareness of the potential problems posed by underwater mateable interfaces, led to some changes of thought in subsea control package design.

For reliability the wells were each provided with a dual redundant high and low pressure hydraulic control umbilical and a separate dual redundant power and signal cable. The major improvement in system reliability was in the way the single combined electrical connector interface was remotely made up directly to the control module, thus eliminating at least one of the former primary underwater mateable interfaces.

The hydraulic supply system however still retained the five underwater mateable interfaces.

Closely following the above development, and driven by economics, control system manufacturers came up with a diver-intensive subsea control module which no longer relied on the traditional expensive vertical entry-base connector.

In the effort to make the hardware more economic by preclusion of the main single combined interface, multiple, diver made up, interfaces were introduced.

FIGURE 2

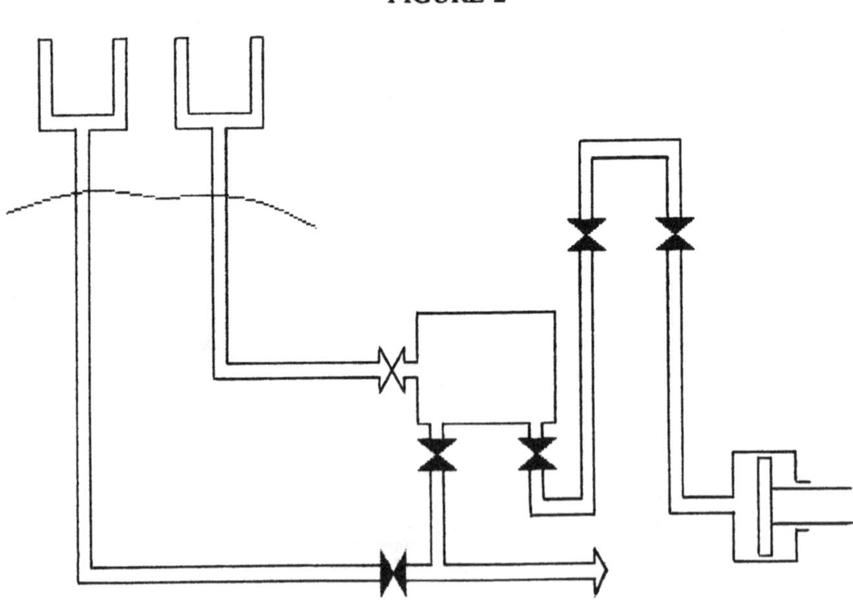

It is considered that the majority of known problems and failures that have occurred on subsea control systems are not generally attributed to failure of discrete items of hardware (always providing that the hardware has been suitably qualified beforehand), but to events involving interfaces. This approach would then seem to have been a retrograde step.

FIGURE 3

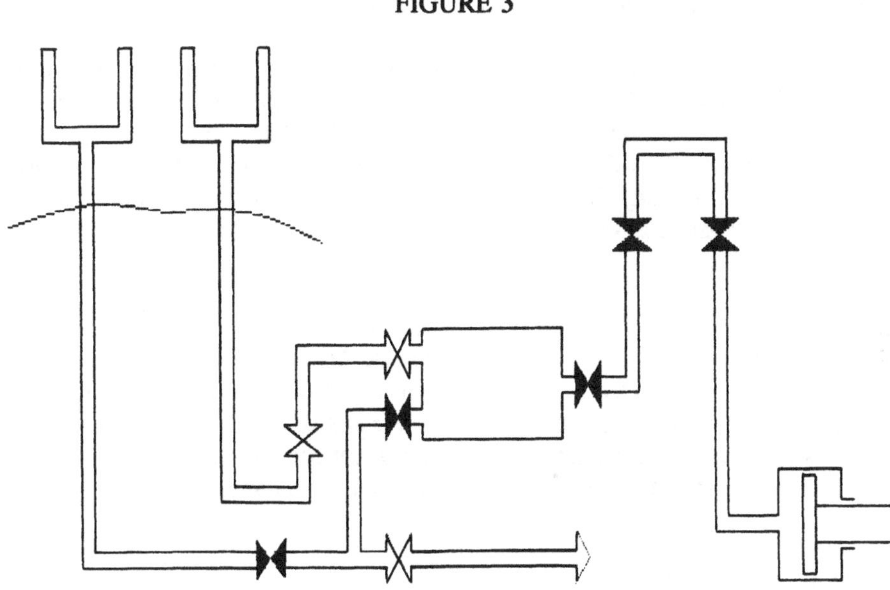

INSTALLATION PRACTICE

Subsea production hardware has traditionally been installed via the moon pool of a semi-submersible drilling rig; the rig being positioned directly over the seabed located guide base structure.

The main process flowlines have almost invariably been towed to a position on the seabed, and subsequently interconnected to the topside process plant and the subsea completion via diver installed spool pieces.

Production control line umbilicals are, in almost all cases, installed at the platform end first and then laid and trenched across the seabed to a target zone close to the subsea hardware. At this stage the umbilical may be temporarily abandoned. The umbilical will lie there until such time as the rig, while installing the subsea production equipment, is able to locate it and with the aid of divers or ROV connect to and pull it into the docking porch on the subsea structure.

During this period of lay-away, abandonment, location and pull-in, it has always been extremely difficult to monitor the umbilical's integrity. It is not until the umbilical is finally connected to the vertically installed part of the system in its entirety that a thorough check on integrity can be made. In parallel with this activity, various other parts of the control system may also have suffered damage or degradation. It can prove to be a costly process to analyze, locate, and rectify any fault or faults.

The production hardware, vertically installed from the surface, normally carries with it the subsea production control package. In order that the equipment is balanced during the descent to the subsea structure, counter weights are fitted to compensate for out of balance mass. A subsea control package can sometimes weigh up to 2,400 kg. A normal bare christmas tree will weigh 25,000 kg. Add to this the 4,800 kg for control package and counter weights and the total is very close to the maximum rig crane capacity on a good day.

Traditionally subsea production control has relied on the provision of hydraulic muscle power supplied from the surface via thermoplastic hoses. These hoses are bundled together and carefully protected by a double steel wire armour sheath. The umbilical is an expensive piece of hardware. The hydraulic power unit supplying fluid to the umbilical normally has just one reservoir. Introduction of random contamination and/or failure to maintain this reservoir in the peak of condition could quite conceivably lead to subsea failures.

THE 'MANTA' CONCEPT.

Without changing basic ground rules too much, the 'MANTA' concept makes it possible to eliminate all of the above difficulties.

In principle, the 'MANTA' concept is based on generating the hydraulic power at the subsea end of the umbilical and retaining the subsea production control functions within the same hard-wired package. Only the actual downstream 'secondary' discrete functions and sensors are fed through the package subsea mateable interface. No primary electrical or hydraulic connections are made up subsea. As a direct result more advance forms of communication and higher levels of electrical power may be transmitted to the subsea package.

The hard wiring of the package is undertaken in a controlled factory environment prior to installation. Thus the overall installed system reliability is greatly improved.

The concept makes provision, as an emergency backup, for the MANTA control package to be disconnected from the umbilical and replaced without causing total flooding of the umbilical lines.

FIGURE 4

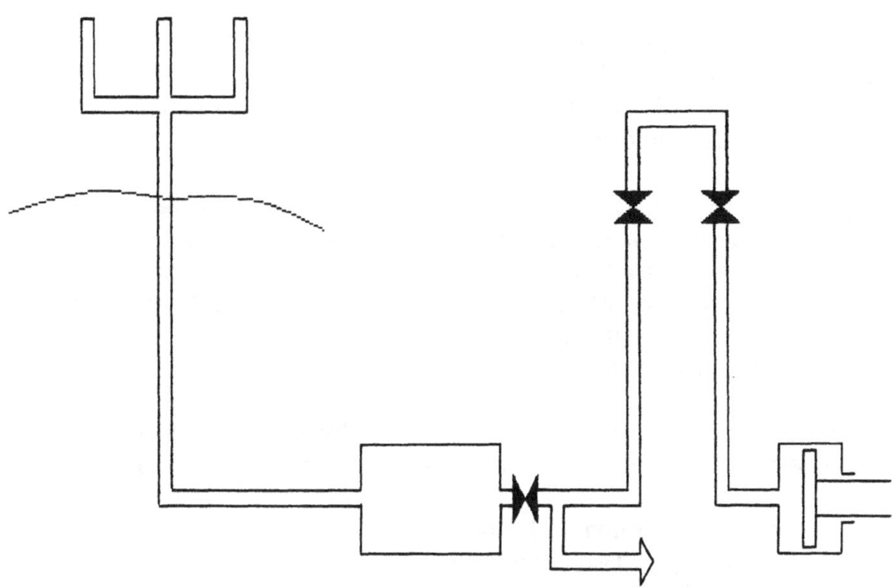

The 'MANTA' concept was originally devised as a way to simplify the subsea end of a production control umbilical. This followed the realisation that any damage sustained by the umbilical during its life would necessitate disconnection and retrieval of one end or the other to the surface. This simple fact has major implications for the reliability of the system, particularly where disconnection and reconnection of primary electrical and hydraulic interfaces would be required. In an attempt to overcome this potential limitation of existing systems the 'MANTA' concept was evolved.

In accepting the fact that, where umbilical damage is sustained, the subsea end may well have to be disconnected from the seabed installation, the case for a very simple subsea control interface is made.

If the criticality ranking of any subsea interface can be reduced to discrete 'secondary' function level only, this, in conjunction with the possibility for ROV/manual override on each function, provides for an order of magnitude improvement in overall system reliability.

By the introduction of these seemingly minor changes the whole control system becomes a far simpler, more attractive and economic package. The advantages enumerated above are discussed in more detail below.

HARDWARE DESCRIPTION.

In the preferred state the 'MANTA' subsea control package is a hydraulic fluid filled, fully compensated umbilical termination. The termination package contains dual redundant electrically driven hydraulic pump units, high pressure fluid intensifiers, multiplex electronics and redundant fluid reservoirs.

Electrical power and communications are provided from the surface via the umbilical and the subsea end is hard wired to the control module internal electrical system. The umbilical incorporates a water block a short distance from the end termination on the control module. Having eliminated the need for any subsea umbilical connectors, it is quite feasible to utilise fibre optic data communications to and from the package. This could include visual capability for installation monitoring.

Hydraulic pressure is generated by two electric motor driven pumps which supply two accumulators with fluid at a pressure suitable for the majority of subsea equipment.

For the downhole control functions, a hydraulic intensifier is employed which is driven from the normal fluid supply. Two piston-type fluid storage accumulators are provided. These are each equipped with a gas back-up bottle mounted externally to the control package structure.

Fluid pressure is directed to the various downstream functions by a set of manifold mounted solenoid operated shear seal valves. These are controlled from the multiplex electronic package.

Discrete hydraulic control functions are taken through a single subsea made up interface at the opposite end of the control module to the umbilical.

The MANTA package is deployed horizontally in line with the general lay of the umbilical. During the last stages of installation, it is hauled across the seabed target zone and into the docking position on the subsea structure. To facilitate horizontal pull-in, the control package is mounted on a punt like sledge. This protects it from any bottom debris and allows it to be easily drawn across the seabed and into the docking porch.

The sledge alleviates any tendency for the package to sink into the mud during the time that it is abandoned after initial laying.

FIG 5

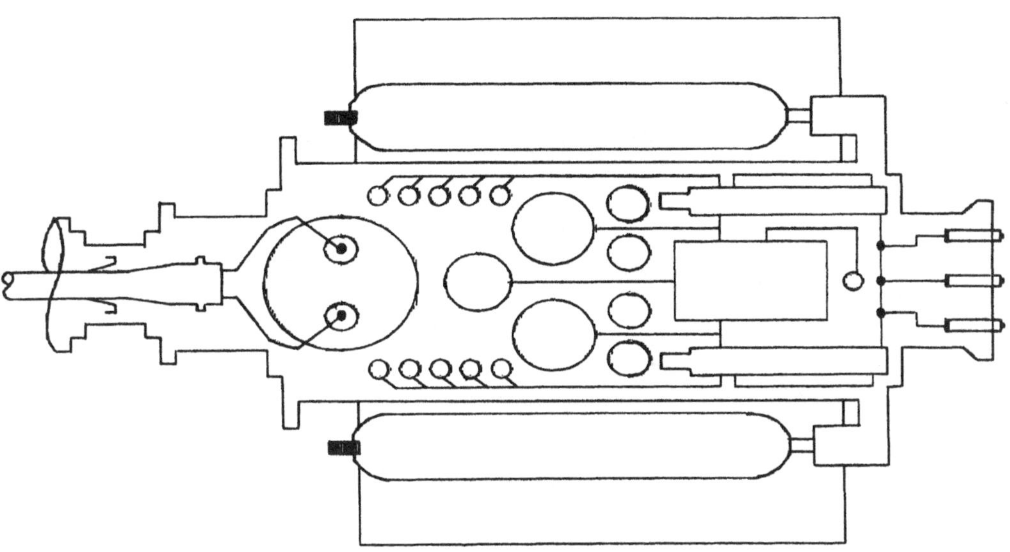

The pull-in mechanism may be of a twin-wire type. This allows for simple steerage and alignment during the pull-in process. It takes the form of a conventional tree mounted system, using the rig's winching facilities. The alternative would be a self contained subsea winch, hard-mounted on the control package. The latter system is considered the better method there being no requirement for the surface 'mother' vessel to have heave compensation or dynamic positioning. This allows the use of a vessel of convenience for the final control system installation.

The control module mounted winch will incorporate the necessary torquing mechanism and tooling for operating the interface clamp during installation, and if necessary, for retrieval of the control package.

On completion of the pull-in and connection the winch package could be detached from the control module and retrieved for further use. The pull-in wires between the winch and christmas tree structure would be attached/detached using a working ROV with manipulators.

FIG 6

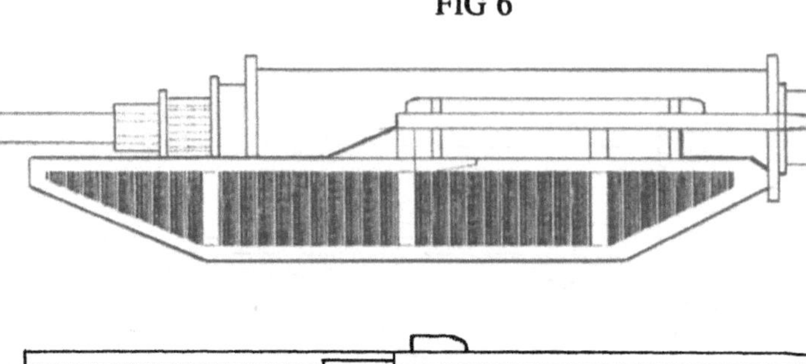

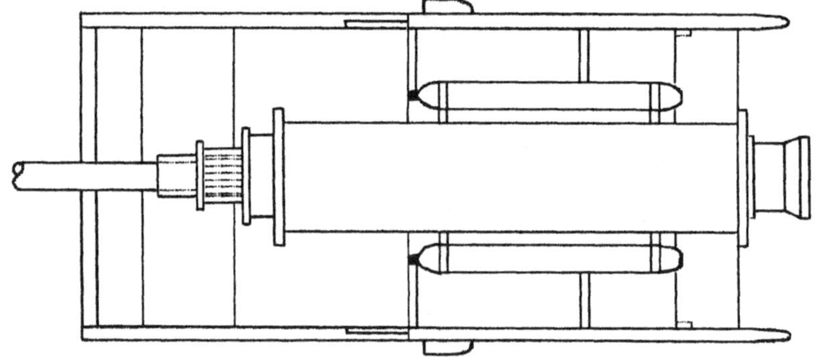

RELIABILITY VERSUS MAINTAINABILITY.

There are a number of schools of thought within the industry regarding the nature of an acceptable subsea control system.

Modularization of equipment is fairly common practice in order to achieve a high level of maintainability. This approach ensures that only equipment directly related to any failure need be removed and replaced. It does not however take into consideration the very real problem of interface failure modes. Clearly the greater the number of interfaces, the greater is the chance of failure.

Modularization as a means of achieving high availability is therefore self defeating.

To maximise reliability in the subsea application a totally integrated and self contained system is perhaps the most advantageous. An example of such an approach may be seen in the development and use of mono-block christmas trees for subsea. The design of the 'MANTA' concept follows such a philosophy.

SIMPLIFIED INTERFACES.

Having the umbilical hard-wired into the back of the control module eliminates a number of headaches which normally present themselves when trying to achieve reliable installation and operation of existing systems. The principle advantage of this approach is the total elimination of the need for subsea make and break primary electrical and hydraulic connections.

With all the control functions occurring within the 'MANTA' module the only connections on the subsea interface are for the downstream control function and status sensors. In real terms, for a single string christmas tree, this means:-

1) Production Master Valve Function

2) Production Wing Valve Function

3) Annulus Master Valve Function.

4) Annulus Wing Valve Function.

5) Chemical Injection Valve Function.

6) Production Pressure Sensor.

7) Annulus Pressure Sensor.

8) Production Temperature.

9) Chemical Injection Line.

10) Common Control Function Return Line.

11) Down Hole Safety Valve Function.

ADVANTAGES.

The concept has a number of major advantages over current standard modular systems.

Firstly - Because of the simplified 'function only' hydraulic interface, the 'MANTA' concept dramatically reduces the overall number of hydraulic lines which need to be made up in the subsea environment. This in turn has a major effect on the number of interconnecting subsea hydraulic lines required for any given number of subsea functions.

FIG 7

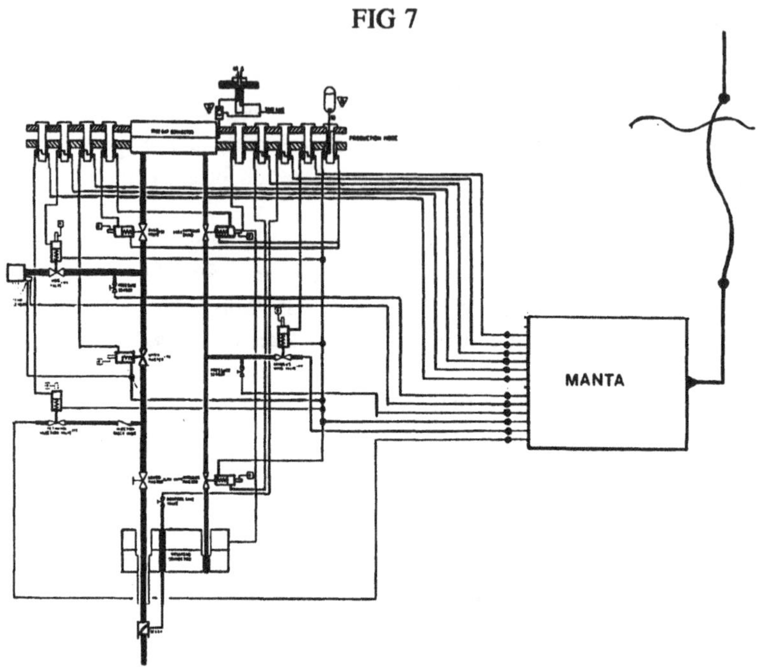

This represents a significant reduction in the number of connections normally made subsea and hence the number of possible failure points.

DECENTRALISED RISK.

Almost all existing subsea production control systems rely for their muscle power on hydraulic power being generated on the platform in a centralised hydraulic power unit. This power unit normally has a common reservoir and sometimes, but not always, incorporates duplicate redundant hydraulic pumps. The introduction of contaminated or incorrect control fluid to this centralised system immediately puts the whole subsea field development at risk. Any common component failure within the centralised hydraulic power unit downstream of the reservoir also puts the whole subsea field development at risk.

The preferred option 'MANTA' ensures that this risk factor is totally eliminated by the utilisation of a discrete reservoir system and dedicated dual redundant hydraulic pumps within each subsea control package. There does however still exist the option to pump fluid from the surface. This could, in practice, be restricted to a back-up system for reservoir make up fluid only.

In keeping with a high reliability approach, it is anticipated that 'MANTA' control line umbilicals would carry redundant electrical power and signal lines.

Decentralising the hydraulic muscle power needed to operate the subsea functions, in combination with the hard wired approach, dramatically improves the total field availability.

Complete failure of any single subsea module would have only a limited detrimental effect on the field. Total failure of any properly qualified decentralised subsea control package is a highly unlikely event in any case. Once the equipment is past its infant mortality stage and has been installed safely and correctly, there should be very little that can actually detrimentally affect it apart from eventual wear out.

It can clearly be seen that, in the case of a centralised hydraulic power unit, any H.P.U. malfunction may have the effect of reducing, or even stopping, the entire field production. However, with several decentralised units such as 'MANTA', only the production directly related to the failed unit will be effected. While a single failure of this nature in itself is highly undesirable, it should be kept in perspective. The probability of the complete failure of a dual redundant system occurring is unlikely.

INSTALLATION INTEGRITY.

Common sense has traditionally dictated that just prior to installing a control module subsea, it should be run through a full test programme to verify total integrity. It has also been shown that some form of testing should be undertaken on the control umbilicals, initially on completion of a 'J' tube pull-up, and then on a routine basis during laying and, if required, trenching, or rock dumping.

It is not always practical or possible to fully test an umbilical in ,its completed state for electrical continuity. Hydraulic integrity is also a very difficult thing to verify, particularly on thermoplastic hoses, as the lines tend to stretch and creep.

With the advent of 'MANTA', it becomes a very much simpler task to initiate and maintain a constant monitoring programme on the umbilical during laying. It also becomes a good deal easier to ensure that the control system will act correctly once on the seabed and ultimately when connected to the subsea hardware.

Despite the fact that there is only a hydraulic interface between the 'MANTA' control system and the christmas tree, process data such as pressures, temperatures, and if necessary flow rates, can be monitored by the control module using existing instrumentation equipment and technology.

PRIMARY ECONOMICS.

The most advanced type of subsea control system at this time may be considered to be of the electro-hydraulic multiplex type.

In order to ensure long term reliability of performance of the interconnecting umbilical, duplication of services should in fact be built in. Currently, and in practical terms, this means that the umbilical would contain:

1) Duplicated electrical power pairs or triples.

2) Perhaps duplicated signal lines.

3) Duplicated low pressure hydraulic supplies.

4) Duplicated high pressure hydraulic supplies.

5) At least a single chemical injection line.

Any reduction in the number of hoses or cables needed, which can maintain the level of reliability of the system, must represent a major improvement in terms of performance and economics.

For the preferred option 'MANTA', the umbilical requirements are restricted to:

1) Duplicated electrical power triples.

2) Duplicated electrical or fibre optic signal lines.

3) A single chemical injection line.

4) A possible single hydraulic fluid make up line.

This configuration represents a very substantial reduction in umbilical cross section and an appreciable saving on materials compared with the umbilical requirements of current control systems. As a consequence, it provides very significant economies in the overall control system budget. The longer the umbilical, the more significant the economy will be.

FIGURE 8

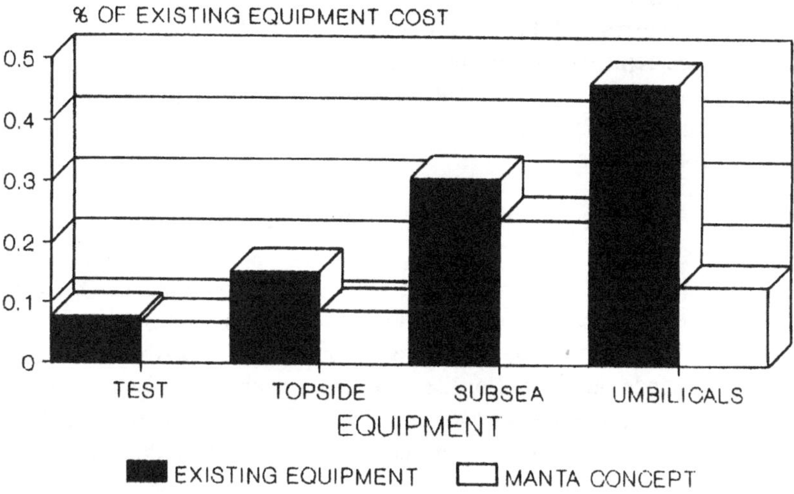

SENSITIVE AREAS.

With any high integrity control system there must always be the questions:-

What happens if....There is an ingress of water into the module or hydraulic system ?

There is a possibility of this happening during connection to the christmas tree interface (as in all wet mateable hydraulic connections). Water can be trapped in the final stages of the mating of the connectors and be forced into the downstream hydraulic circuit. Eventually this water will find its way back to the main reservoir in the control module (due to fluid movement during actuator operation) where it will sink to the lowest point of the module. By discharging the accumulators to the return system, and thence back to the reservoir, the water will be displaced and ejected through low level relief valves. This also applies to water ingression from other sources.

What happens if....Sour gas becomes present in the return line from the D.H.S.V. ?

Any gas present in the return line, from the D.H.S.V. or any other source, will be separated out and vented to sea by the in-line gas vent valve.

What happens if.....There is loss of hydraulic fluid due to D.H.S.V. leakage ?

Where there are fluid losses due to external leakage of any kind, the loss will be initially made up by expansion of the reservoir compensators. Fluid level indication incorporated within the compensators will provide early warning and, at the appropriate time, the reservoirs can be replenished. This may be directly from the platform through the umbilical (requiring an extra umbilical hose) or, from a vessel of convenience positioned locally on the surface replenishing the reservoirs via a hose connected to the control package by a ROV or diver. Alternatively a flexible container can be lowered to the seabed adjacent to the subsea installation, and connected by a ROV/diver again making use of a vessel of convenience.

What happens if.....There is a total control module failure ?

In the event of a total failure the control module will be unlatched using a ROV/ROT and jacked away from the interface until clear of the guide funnels. Depending on just how well the umbilical has been protected, and how accessible it is, the MANTA package may be retrieved to the surface on the end of the umbilical, or alternatively cut free just aft of the bend restrictor and again just forward of the water block. The control package and the main umbilical upstream of the water block may then be pulled to the surface separately.

The repaired or replaced control package, which if necessary would be supplied with a new short length of umbilical, would be redeployed and pulled in and connected to the subsea structure as in the initial installation scenario.

What happens if.....The christmas tree has to be removed ?

Once the well has been rendered safe the control module may be unlatched and jacked clear of the interface. The christmas tree can then be removed, a replacement fitted, and the control package replaced. Throughout the operation full communication can be maintained with the control package.

These are but a few of the "what happens if" questions but, if there are any others, it is hoped that the information contained in this paper is sufficient to answer at least a few of them.

SUMMARY.

The 'MANTA' control module is designed as an integral part of the submarine umbilical with only the essential minimum number of hydraulic functions being taken through the subsea interface. This approach eliminates the need for primary electrical and hydraulic connections to be made up subsea.

The MANTA concept provides major improvement in system reliability and a substantial reduction in system costs.

By making provision for a hard wired primary interface, and with the extra electrical power that may then be transmitted, hydraulic power generation is achieved subsea. This dramatically reduces the essential umbilical services required.

The complete subsea control system right downstream to the secondary subsea interface can be commissioned prior to deployment and continuously monitored during lay away from the topside facility.

Once the christmas tree has been installed, the rig can be released for other work and a more economical vessel used for the connection of the control system.

As the subsea control module is deployed horizontally there is no need for excessive christmas tree counterweights, thus dramatically reducing the load out and handling weight of the production equipment.

In conclusion, very considerable cost savings can be achieved, particulary for remote marginal fields.

It is hoped that this paper has given some food for thought or at least provided the basis for questioning traditions that have limited foundations.

Session 2
Technical Innovation

BUILDING IN COST SAVING - DIVER RECONNECTABLE UMBILICAL WEAK LINK PROTECTION SYSTEM

M.G. WARREN
Subsea Project Engineer,
Offshore Engineering Division,
Strachan & Henshaw Ltd.

ABSTRACT Hydraulic, Electrohydraulic and Multifunction umbilicals form a critical and key component in Subsea Oil and Gas production systems. A recent Government report has documented evidence which supports the longstanding concern of operators, contractors and regulators over their reliability.

Among recommendations made in that report was the need for improvements in overload protection to prevent expensive damage caused by trawler and ship anchor snagging.

This report documents the development of a new generation of 'weak link' connector, its use in current and future subsea systems and the associated savings both in capital and operational terms.

1.0 INTRODUCTION

With the increasing need to develop marginal fields, the use of subsea oil and gas production systems have increased over the last 20 years quite dramatically. There are currently over 150 subsea completions in the North Sea alone, with some predictions estimating that by the year 2000 this will rise to 400. The North Sea is in the forefront of this development, with its relatively shallow waters making satellite well tie back to existing production platforms that much easier. The past decade has seen the trend for the use of more satellite wells located at greater step out distances. In addition, the complexity of subsea control systems has increased, demanding higher reliability of critical components. One of these critical components is the umbilical.

Volume 30: Subsea International '93, 69–92.

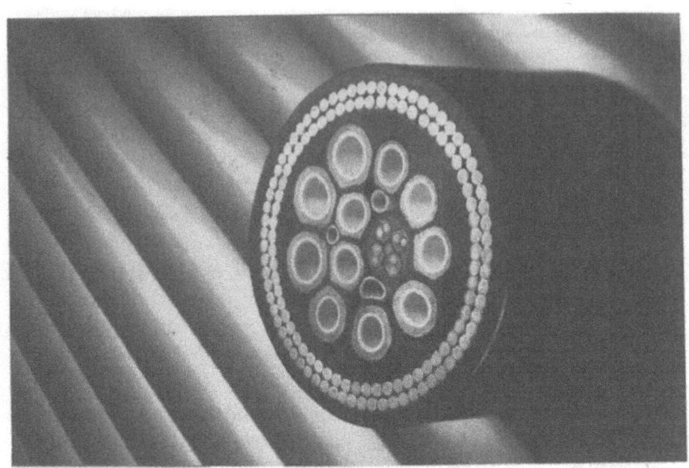

FIG.1
UMBILICAL TYPICAL CROSS SECTION

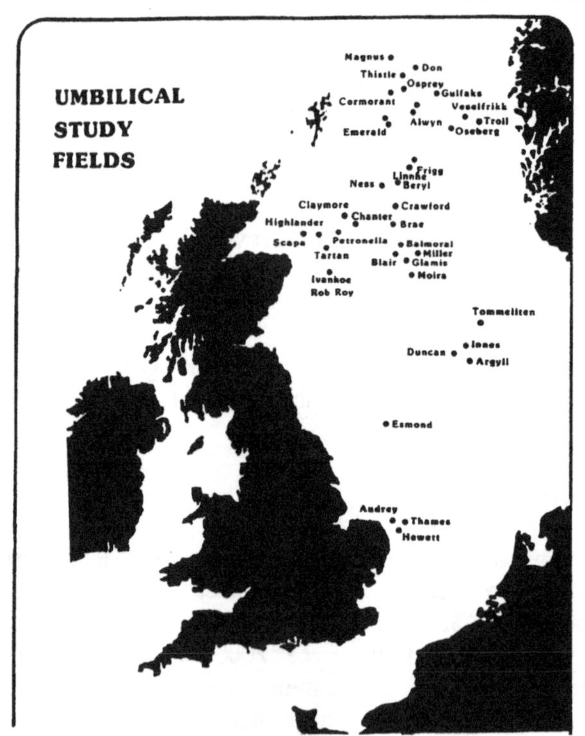

FIG.2
ERC FIELD SAMPLE AREA

The subsea umbilical performs a multiplicity of functions which can be summarised as follows:-

- Control of subsea-completed production or injection wells.
- Control of pipeline safety valves.
- Injection of chemicals in subsea wells.
- The Distribution of electrical power.

Fig.1 shows a typical umbilical cross section.

Operators, contractors and regulators are therefore demanding umbilicals which are longer, more sophisticated and more reliable than ever before. The umbilical is however, a very vulnerable key component, which is exposed to the high risk of significant external mechanical damage.

A recent survey by the Engineering Research Council for the Department of Energy showed that from a total selection of 180 umbilicals, 54% had suffered from various problems. Fig.2 shows the areas from which these samples were taken. Out of the samples analysed, just over 15% of umbilicals which suffered from mechanical damage during service received that damage as a result of trawler and anchor chain snagging. Whilst in overall numbers, the actual amount of recorded incidents is relatively low, the possible effects can be catastrophic.

A typical armoured umbilical could easily sustain an axial tension of 60 tonnes without breakage. This could inflict severe damage on its interface and associated equipment, such as a Tree or Manifold for example.

To protect critical equipment and the umbilical itself, some operators incorporate a 'weak link' in the umbilical line. This weak link usually takes the form of a mechanical device designed to separate in a controlled manner at a pre-set axial overload force. It is located in a variety of positions such as the bottom of the J-tube, or some distance from the platform/well-head base. More than one unit is often incorporated in a single umbilical.

In some systems, the umbilical anchorages are designed to withstand loads higher than the actual umbilical breaking point, making the umbilical the weak link.

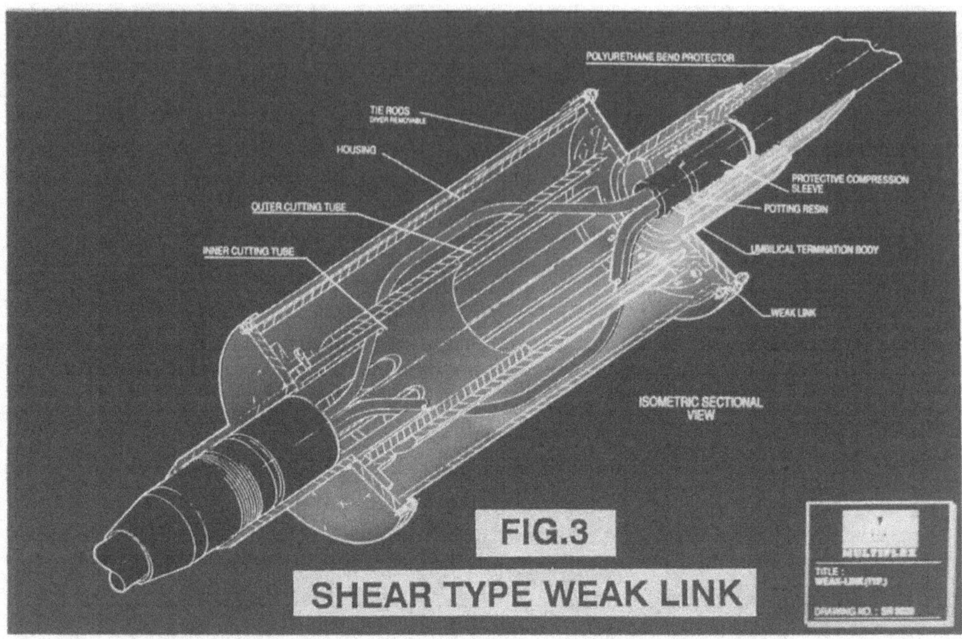

FIG.3

SHEAR TYPE WEAK LINK

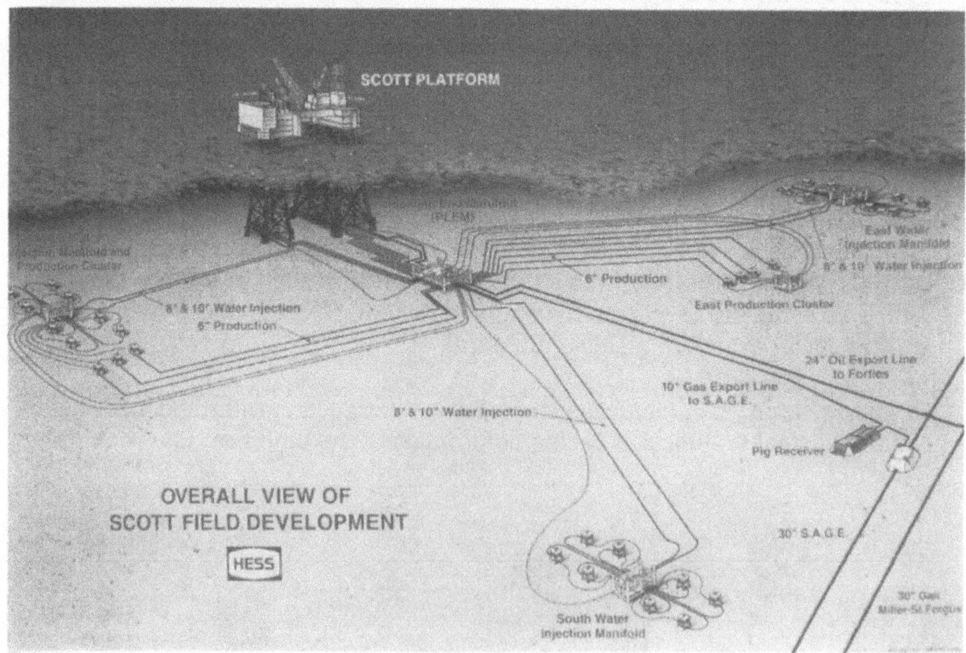

FIG.4

TYPICAL FIELD LAYOUT

2.0 CURRENT METHODS OF OVERLOAD PROTECTION

Existing devices work by shearing the individual lines within the umbilical when axial overload is experienced. Fig.3 shows such a typical weak link. This is achieved by passing each line through matched holes between an inner and outer tube. Each tube is attached to a different half of the umbilical through the outer armouring. The inner and outer tubes are held together by the use of necked bolts designed to break at a predetermined load.

In an overload scenario, these weak link bolts break, allowing the inner and outer tubes to separate. This causes the individual umbilical lines to shear.

There are variations on this design, but all follow the same general principle. The results of this type of unit being activated can be summarised as follows:-

- Line contamination (i.e. seawater ingress).
- Cut umbilical lines.
- Local pollution from line spillage (i.e. chemical loss).

In addition to these 'proprietary' weak link units, other methods are employed to achieve the same effect. For instance, tree protection can be provided by the use of a horseshoe type of design. This makes use of a U-shaped umbilical termination joint bolted to the tree interface with shear bolts. The individual hoses are of varying lengths and terminate in hydraulic fittings. In the event that the umbilical becomes axially overloaded, the shear bolts break, allowing each hose to pull off its fitting at the swaged joint. The use of varying lengths of hose ensures that tension is transmitted through one hose at a time, thus keeping the overall tension to a minimum.

This type of weak link termination incurs less overall damage to the umbilical, but still gives rise to line contamination and fluid spillage.

The ERC report suggests that this traditional approach of cutting the umbilical should be reviewed, with self-sealing hydraulic connectors being used.

3.0 THE WEAK LINK DEVELOPED BY STRACHAN AND HENSHAW

3.1 Development

The Strachan & Henshaw Weak Link Connector System was developed for use on Amerada Hess' Scott Project. The Scott Field is located in 460 feet of water in the Outer Moray Firth area of the North Sea, 130 miles North-East of Aberdeen. The subsea layout (Fig.4) incorporates a chemical injection system and uses a series of umbilicals to distribute chemical injection fluid to the various wells.

The Strachan & Henshaw Weak Link was incorporated in two of these umbilicals.

3.2 Design

3.2.1 Specification

The requirements for the Weak Link Connector for the Scott field were as follows:-

- Number of lines - 15 x ½".
- Type of fluid - chemical injection (methanol/xylene).
- Pressure (WP) - 7500 psi.
- Pressure (TP) - 10,000 psi.
- Type of Connectors - Hydraulic self sealing fitted with metal seals.
- Water Depth - 460 ft.
- Disconnection Load - 4-5 Ton.
- Design Life - 20 years minimum.
- Umbilical Connection Method - Flanged.

3.2.2 Design Philosophy

The existing shear type weak link connectors described in Section 2 have a very simple method of disconnection viz, the failure of 'weak link' bolts. The wasted bolts are machined with a reduced diameter sized to break at the desired break-away load. The device has no internal forces as the hoses contain the system pressure.

Once hydraulic connectors are introduced into the system however, reaction forces are created when they are under pressure, this tends to force both halves apart. This force is a function of pressure multiplied by effective cross-sectional area less seal friction. A typical hydraulic connector unit as specified for use on the Scott Project would produce about 5400 lbf (2.4 tons) at working pressure. If this separation force is multiplied by the total number of connectors used, it can be seen that the overall separation force will be approximately 81,000 lbf (36 tons).

If the conventional method of 'weak link' bolts is employed in this situation, the bolts will have to be sized to withstand the hydraulic reaction forces and plus the desired separation force. This could be in the region of 41 tons. In addition, the system could only be effective as a weak link when all lines are fully pressurised.

The Strachan & Henshaw engineered Weak Link Connector (see Fig.5) overcomes this problem by isolating the disconnection system forces from the hydraulic reaction forces. The following description shows how this is done by the use of a simple mechanical system:

3.2.3 Description

The Strachan & Henshaw Weak Link Connector has at its heart, two matching circular stabplates each housing a series of hydraulic connector halves (Fig.6). One stabplate houses the male halves, while the other contains the female halves. The plates are profile recessed to allow them to come face to face around their edges.

The stabplates are each profiled with a 45 degree chamfered edge and are held together in a clamped position by a series of externally mounted collets (Fig.7). These collets have a matching internal 45 degree profile and are held in position by an outer retaining sleeve. This sliding sleeve has a slight taper on its inside diameter which matches the outside profile of each collet.

One side of the unit comprises of a tube which houses the umbilical lines and terminates in an end flange for the umbilical connection. This tube is welded to the rear of one of the stabplates and incorporates access windows for the hydraulic connections between the hose fittings and the hydraulic connectors.

The other half of the weak link incorporates the disconnection mechanism (Fig.8). This comprises a short profiled tube welded to the stabplate running inside a larger outer sleeve, which also has access windows and an umbilical termination flange.

FIG.5

STRACHAN & HENSHAW WEAK LINK

FIG.6

S & H WEAK LINK STABPLATE

FIG.7
DETAIL OF COLLETS

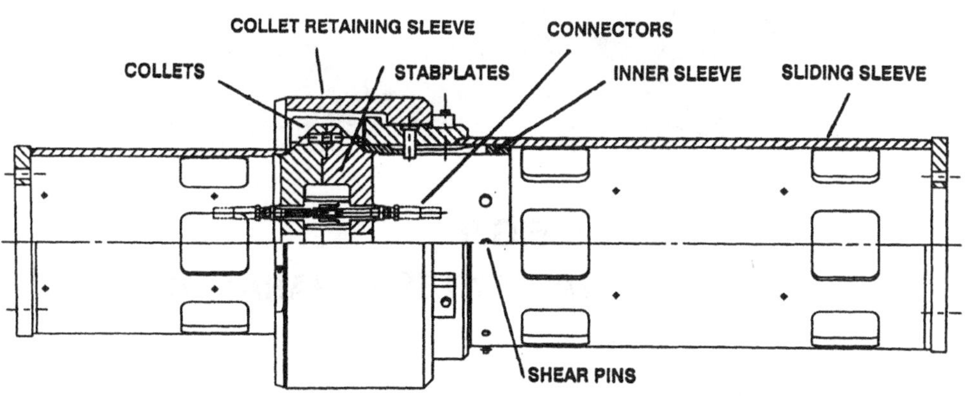

FIG.8
S & H WEAK LINK CROSS SECTION

FIG.9
DETAIL OF STABPLATES SEPARATING

The outer sleeve is held in a fixed position by 3 shear pins radially passing between both inner and outer tubes and is indirectly linked to the collet retaining sleeve by a lost motion joint.

Upon activation, the shear pins break, allowing the outer sleeve to slide along the body until it makes contact with the collet retaining sleeve. Up to this point, the stabplates remain in contact and the hydraulic connectors remain fully engaged. After contact is made, the collet retaining sleeve is pulled back, exposing the collets which are then released, allowing the stabplates to separate (Fig.9).

The travel of the various sliding components is limited by mechanical stops to ensure the collets are retained within the unit upon disconnection, ensuring that they do not become lost on the seabed.

Sufficient additional length of umbilical hose is incorporated within the weak link body to accommodate the lost motion mechanism and outer tube travel.

Extensive development work was required to achieve the operations described as initial calculations had demonstrated that the internal friction between collets, stabplates and the retaining sleeve was very high which would have resulted in the mechanical disconnection forces being in excess of the required disconnection load.

3.2.3 Hydraulic Connectors

The hydraulic connectors used in the Scott field application were the self-sealing metal to metal type (Fig.10) and incorporates a pressure energised metal seal, and a Viton back-up seal.

Where line bleed down is required on disconnection (for example, on a subsea valve hydraulic control line), the poppets are drilled through with a small hole to allow the fluid in the downstream side to vent out.

The Strachan & Henshaw Weak Link system can accommodate a variety of hydraulic connectors and electrical connectors.

3.4 Manufacture

The Strachan & Henshaw Weak Link was manufactured using conventional techniques and a variety of material with various protective coatings.

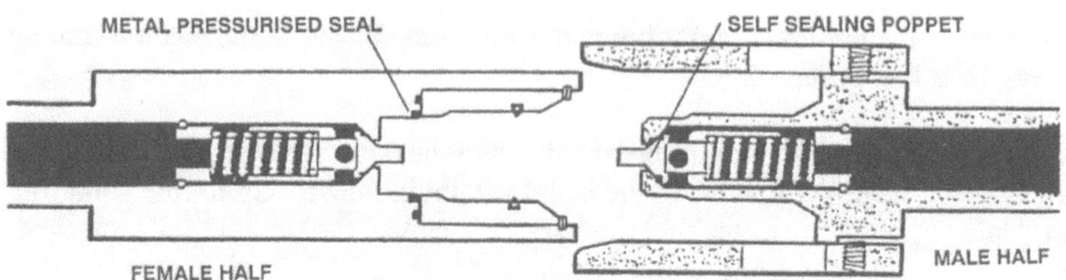

FIG.10

SECTION THROUGH TYPICAL HYDRAULIC COUPLER

FIG.11

WEAK LINK DURING TESTING

The hydraulic connectors, stabplates and stabplate tube body were all manufactured from stainless steel. Other components such as the outer retaining sleeve, sliding sleeve, collets etc., were manufactured from a high grade carbon steel and surface coated with a subsea quality 4-coat epoxy paint applied by airless spraying. Sliding surfaces were coated in a low friction PTFE finish and contact areas between mating components were all unpainted to ensure electrolytic contact which is a requirement of the cathodic protection system.

3.5 Testing

Testing was performed in the following stages:

- Proof pressure testing.
- Functional testing.
- Shear pin testing.
- Combined pressure/functional testing.

3.5.1 Proof Pressure Testing

This was carried out in two stages.

- Separated.
- Connected.

The separated halves of the weak link were individually pressure tested to prove the sealing integrity of each hydraulic connector. This was achieved by manifolding all the male halves together in one half and all the female halves together in the other. This test demonstrated that each self-sealing poppet was working efficiently.

Another pressure test was performed with the two halves of the weak link connected to prove the sealing integrity of the main connector seal. This was achieved by manifolding all the male halves together and fitting blanking plugs to hoses running from the backs of the female connector halves (see Fig.11).

The blanking plugs were initially loosened to not only bleed air from the system, but also to demonstrate that there was an unrestricted flow of hydraulic fluid. This is particularly important to note, as it is quite possible to successfully pressure test a system with the flow being either severely restricted or even shut off. Reports have attributed subsea failures to this very situation, where insufficient engagement of hydraulic self-sealing connectors leaves the poppets either fully closed or insufficiently opened. Insufficient opening is a particularly dangerous situation because an inspector can be deceived into concluding that all is well when he has seen the witness of fluid which has passed through the connector.

What can then happen is one of two things (or a combination of both):

i) Hydraulic reaction between the connector halves can move the connectors further apart, reducing the poppet clearance even more or effecting total closure. A large number of connectors in a stabplate impart a very large separating force, as shown in Table 1. These combined forces can even deflect a thick stabplate by a significant amount relative to the small poppet stroke.

ii) Insufficient opening can allow the poppets to float in response to fluid movement. This can cause intermittent closure of the poppets or even total line shutdown.

Hydraulic System Pressure Bar (psi)	Reaction Force N(lbf) per Hydraulic Connector
204 (3000)	9607 (2160)
340 (5000)	10,013 (3600)
510 (7500)	24,019 (5400)
680 (10,000)	32,026 (7200)

TABLE 1

3.5.2 Functional Testing

Prior to a fully pressurised operational simulation, an unpressurised functional test sequence was performed. This ensured that all aspects of the mechanical operation during weak link activation was satisfactory.

The benefits of performing this test were as follows:-

- Slow motion disengagement - visual monitoring.
- The ability to measure collet clearances etc. at intermediate stages of disengagement.

3.5.3 Shear Pin Testing

As previously discussed, the shear pins are the critical components which must break at a predetermined load prior to mechanical operation and disconnection of the weak link system.

Previous experience has shown that shear value results from actual testing are not consistent with calculated values. Therefore, a range of shear pins were manufactured with varying groove root diameters and a short testing programme was undertaken to obtain the desired results. This was done using a test rig which simulated the actual equipment conditions. To ensure consistency, acceptance was based on successful testing of three identical pins.

3.5.4 Combined Testing

Actual operating simulation was performed by a combined test. This test consisted of physically pulling the two halves of the weak link apart with the shear pins fitted and the hydraulic system fully pressurised.

A dedicated test rig was anchored to the floor to which one half of the weak link was bolted and the other half was pulled upwards through an adaptor by an overhead crane. The vertical pull was monitored using a calibrated statometer.

Fig.11 shows the test rig and weak link assembly.

FIG.12
TYPICAL UMBILICAL DURING INSTALLATION

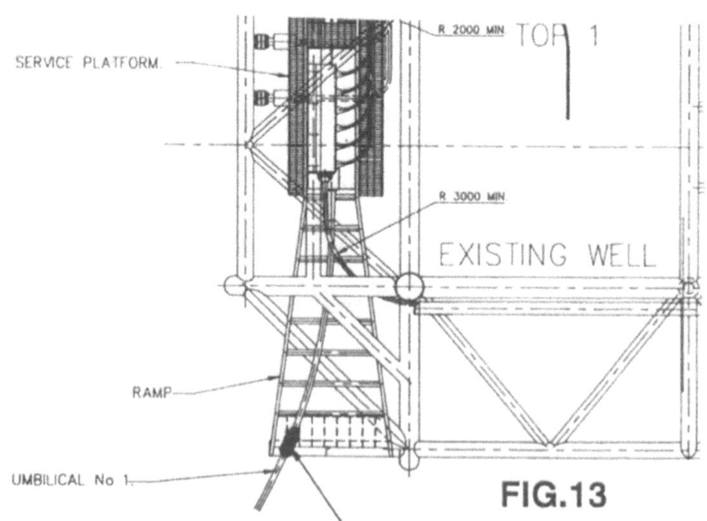

FIG.13
WEAK LINK IN TYPICAL POSITION

3.6 Test Results

The two tested units actuated within 0.25 of a tonne of each other at 4.75 and 5.00 tonnes respectively. This was due to the shear pin setting with the subsequent mechanical operation/disconnection operating at under 1 tonne. Disconnection occurred very quickly with minimal line fluid spillage. In actual fact, the hydraulic system retained nearly 50% of its working pressure after disconnection.

No physical damage occurred to any of the weak link components over the testing programme, including the coated surfaces at the collet/retaining ring area.

3.7 Installation and Reconnection

3.7.1 Installation

The weak link is designed to be installed as part of the umbilical in its made up condition. Fig.12 shows a typical umbilical prior to installation. During topside handling and installation safety pins are incorporated to ensure that the unit is not accidentally activated by inadvertent overload.

Incidents have been reported where umbilical damage has been incurred due to axial overload during the installation/unreeling operation. To prevent this, a higher rated set of shear pins can be fitted for the installation operation and subsequently replaced with the operational in service pins later. These safety pins (or higher rated shear pins) are removed either by diver or ROV after installation.

3.7.2 Reconnection

After an emergency disconnection has taken place, the following steps are required to reconnect the unit:

- Locate the separated umbilical/weak link ends.
- Inspect the umbilical for damage.
- Pull both halves together.
- Clamp stabplates together.
- Pull on collet retaining sleeve.
- Replace shear pins.

The first task is to establish the cause of the separations, locate the separated ends of the umbilical/weak link halves and inspect for any mechanical damage.

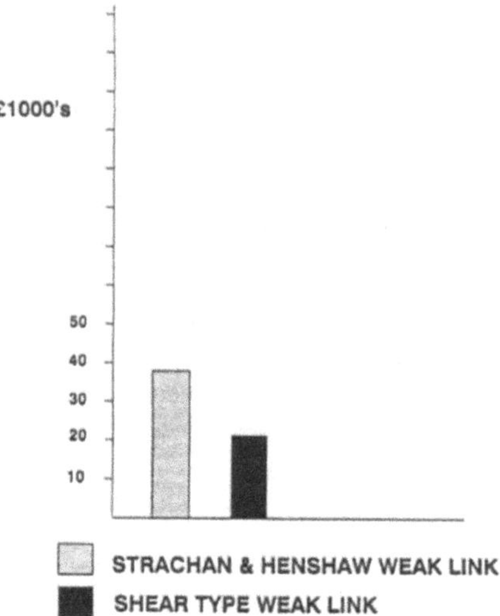

EQUIPMENT CAPITAL COST
FIG.14

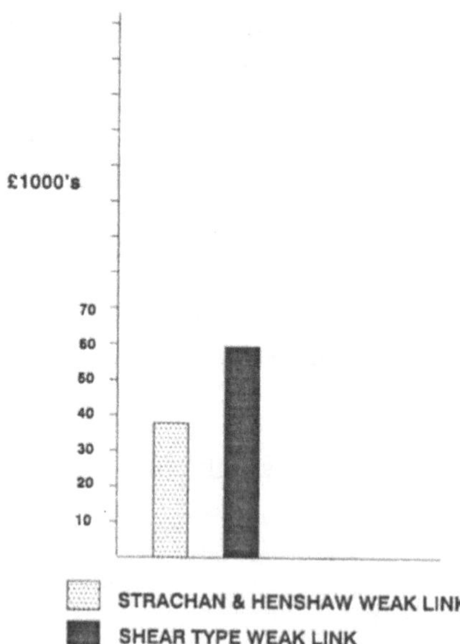

EQUIPMENT CAPITAL COST
FIG.15

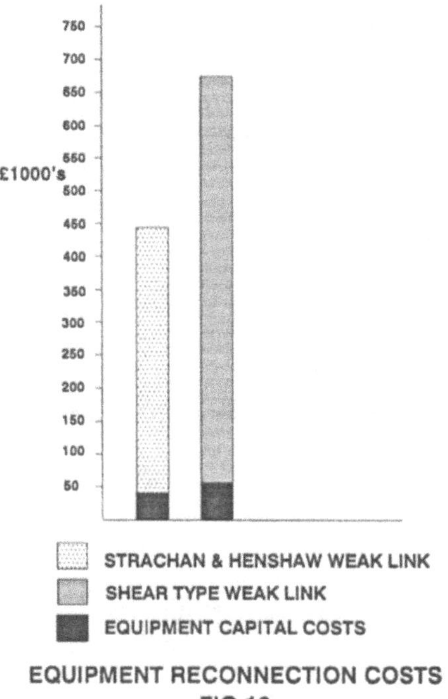

EQUIPMENT RECONNECTION COSTS
FIG.16

FIG.17
ARTISTS IMPRESSION OF WEAK LINK SUBSEA

Conventional winching and pull-in techniques are then used to pull the separated halves into close vicinity with each other. A protection cover/bull nose is recommended during this operation.

The next operation is to clamp both stabplates face to face. This is carried out by the use of conventional studs/nuts and can be performed by hand or hydraulically. This operation avoids the use of integral screw threads in the flanges as marine growth etc. would make them unworkable after a short period on the seabed.

After this operation the collet retaining sleeve is pulled into position by a similar method and after successful remating, new shear pins are fitted. These are designed so that a diver can insert them and are retained large spring pins. Fig.17 shows an artist's impression of the Weak Link in its subsea environment.

The use of an ROV to perform reconnection is discussed in Section 3.8.2.

3.8 Developments

Since the equipment was designed and successfully installed in the Scott field, further developments in its use and application have taken place.

3.8.1 Optimisation of Reconnection Methods

Due to specific design requirements, the original units proved awkward to reconnect by a diver. This was mainly due to insufficient radial location for the collets. Also, because the clamping studs were inserted through the collet interspaces, unnecessary dexterity was required from the diver.

To overcome these minor problems, the following modifications have been adopted:-

- Collet radial support (i.e. spacers).
- External clamping lugs.

3.8.2 ROV Reconnection

For diver restricted areas and for deeper waters, the use of ROVs is widespread and as their capacities, performance and range of activities are ever increasing, there is correspondingly now a need to enable R.O.V. reconnection of the Weak Link.

From experience gained on previous projects in which the use of ROVs to install umbilical flying lead connectors had been designed and successfully used, an ROV system for weak link reconnection has been developed. The equipment for reconnection is then as follows:-

- The weak link halves are fitted with ROV 'friendly' towing/winching points for the repositioning operation.

- After the two halves are repositioned, a small alignment and clamping cradle is lowered over the units. This cradle locates on alignment and reaction lugs which are fitted to the weak link outer profile.

- The cradle is provided with a series of hydraulic cylinders which perform the following functions:-

  - Pull the stabplates face to face.
  - Pull the collet retaining sleeve into position.

Shear pin replacement is a slightly more difficult task, but can be performed by prepositioning the new shear pins in the correct radial position in the cradle and be inserted by use of small cylinders. A directional spring clip in the shear pin prevents the pin from disengaging while in service.

3.8.3 Using the Weak Link as an Umbilical End Termination

So far, the weak link has been discussed in terms of its application use at some position in the umbilical line. However, as the unit is basically a stabplate connection system, it can be used as a combined umbilical termination and weak link unit.

One of the requirements of the weak link connector is that substantial umbilical overloading must be applied in an axial direction. When the unit is incorporated in the umbilical, this does not present a problem due to its freedom of movement. To allow axial loading to occur when the unit is used as a termination assembly, the freedom to self align with the load must be maintained.

One method of achieving this is to provide a swivel or vertical hinge in the fixed structure. The mobile umbilical based weak link half can then be installed. This method also has the advantage of allowing the umbilical to be pulled in over a wider angle range that a fixed stabplate termination.

Alternatively the weak link can be located just outboard of the subsea structure and anchored back with chains. A short length of umbilical or the individual lines can then be connected into the manifold. The chains will allow the weak link to orientate to the direction of pull. A typical layout of this is shown in Fig.13.

3.8.4 Electrical Connectors

The Department of Energy report recommends that inductive couplers should be used in the new generation of weak link connectors.

The Strachan & Henshaw Weak Link can incorporate a wide range of both inductive and conductive connectors.

4.0 COST

4.1 Introduction

The use of weak link connectors is accepted in the industry as necessary to protect critical subsea equipment. The benefit to safety and cost can be clearly seen in the analysis of accident scenarios, but it is not the purpose of this technical paper to justify the use of weak links on these grounds.

A comparison can be made however, between the costs associated with the traditional shear type weak link and the new generation Strachan & Henshaw Weak Link.

4.2 Capital Cost

Although the design is very simple, this new generation of weak link is initially more expensive than the shear type. The relative capital costs are shown in Fig.14. If the unit is used as an umbilical termination connector, then the capital costs differences are negligible.

However, some operators insist that additional umbilical length is incorporated in the system to allow the umbilical to be recovered to the surface in the event of a Weak Link activating. This additional length is not required when a Weak Link is repaired in situ on the seabed. If this extra umbilical is included in the capital cost, the benefits are clear. Fig.15 shows the effect of this based on a typical umbilical cost of £200 per metre and a water depth of 200 metres.

4.3 Operational/Reconnection Costs

Each umbilical location and potential damage scenario is unique. It is not therefore possible to quote specific costs for weak link reconnection, but as an indication. Fig.16 shows the relative costs where a typical diving vessel and a diving spread are utilised. The chart is based on a 5-day mobilisation charge of £60k and £20k per day respectively.

The relative costs are based on the following:-

STRACHAN & HENSHAW WEAK LINK CONNECTOR		SHEAR TYPE WEAK LINK CONNECTOR	
1.	Diver reconnection subsea.	1.	Umbilical end raised to surface.*
		2.	Repair joint.
		3.	Umbilical re- deployment.
		4.	System purging.

TABLE 2

* NOTE:- This scenario assumes that there is sufficient umbilical length subsea to allow retrieval to take place. This may not be the case in all situations and total umbilical removal may be necessary, which will further increase costs.

As can be seen from Table 2, the shear type unit requires much more work and effort to reconnect. It has been estimated that these activities will take several days extra to perform over and above that required for the Strachan & Henshaw Weak Link system. This is reflected in the cost comparison shown in Fig.16.

The use of an ROV can reduce reconnection costs even further.

Conclusion

The technical innovation and developments described in this paper have resulted in a new generation of Weak Link Connector which offers many distinct advantages over existing equipment. Its ability to be reconnected subsea by a diver or an ROV reduces downtime and costs significantly. In addition to operational cost savings, capital costs are also reduced where the equipment leads to the requirement for a shorter umbilical.

REFERENCES

1. Knight, P.M., O'Mahony, S.G. and Ricketts, D.C., Dept. of Energy (1990) "Study of the Performance and Reliability of Hydraulic, Electrohydraulic and Multi-Function Umbilicals".

2. Figs 1, 3 and 12 by courtesy of MULTIFLEX INC.
Fig.4 by courtesy of AMERADA HESS.

ACKNOWLEDGEMENTS

Amerada Hess
Multiflex Inc.
Weir Group plc
National Couplings

SUBSEA METERING FOR FISCAL, ALLOCATION AND WELL TEST APPLICATIONS

W G EDWARDS
Subsea and Pipelines Division
Kvaerner H&G Offshore Ltd
Davis House 69-77 Robert Street, High Street, Croydon, CRO OYA
United Kingdom

ABSTRACT Well testing of subsea step out fields can be more economic and effective if a metering unit is installed near the wellhead. Such a unit has been developed from the BOET Subsea Separation programme and uses control system elements already certified and demonstrated offshore. It is developed for small gas fields and comprises a single separation vessel with ultrasonic and Coriolis metering principles for the separated gas and liquids. The unit layout is based on the size, weight and guidebase installation of a typical subsea Xmas tree.

INTRODUCTION

Many future UKCS gas developments are likely to be small isolated accumulations. These may be developed as subsea well clusters, with the production commingled at a subsea manifold and transported by infield pipeline to a host platform.

For accurate and regular well testing a second test pipeline will be needed with associated platform riser and test separator. These are costly facilities, particularly the pipeline, which can be avoided if a test separation and metering unit can be located on the seabed at the subsea field instead of being on the platform topsides.

Volume 30: Subsea International '93, 93–108.

This technical paper describes a subsea metering unit based on the BOET subsea separation experience. It will be located, as a seabed installation, in the export flowline route from either a single subsea gas well, or a subsea test header in the case of a manifolded cluster.

The unit is intended to measure the flow rate of hydrocarbon gas and liquids and any associated water, to well test accuracy of about 5%. It is designed and developed with the potential to measure to allocation metering accuracy of 2-3%, after 6-18 months field experience.

The first unit in service will therefore be for well test with the option that it can be uprated in service to allocation metering duty as the gas field, and adjacent fields are developed further. It will also be the first step in the longer term development of a subsea fiscal meter for gas wells. Fiscal metering will require 1% accuracy and a higher standard of self calibration and sampling to meet gas customer and government requirements. The development of such a fiscal meter is proposed as a longer term programme, based on the service experience of the well test/allocation meter.

In principle the unit will take the form of a single stage gas/condensate separator with flow measurement systems in the gas and liquid pipework. After measurement (and sampling where necessary) the product phases are recombined in the export flowline (although they could remain separate if the extra flowline for liquids is justified). The unit is designed to fit on a standard AP1 4-post guidebase either in a template wellbay or as an isolated seabed unit. It will usually be supported on 30 inch casing piled into the seabed and run and secured by normal wellhead equipment and procedures. Autonomous control is provided continuously by a subsea PLC (programmable logic controller) in a dry subsea module. Power and services are provided by a short electrohydraulic umbilical jumper from the subsea termination of the main umbilical which serves the trees and manifold. The unit is referred to in this paper as a phase splitter and metering unit or PSM.

The simplest application is shown in Figure 1. The PSM will be installed near a subsea well during the same jack-up operation and therefore within the drilling envelope of the derrick. If several cluster wells are considered the PSM will be on the subsea manifold structure.

The basic configuration is shown in Figure 2. It will be a simple diver installed unit which can be retrieved at any time by a small diving support vessel (DSV). The subsea control module can be retrieved and redeployed as a separate operation by a work ROV, (remotely operated vehicle) without the need for divers.

The design is basic and simple with diver connected spools and umbilical.

The design can also be enhanced to provide diverless recovery/redeployment of the unit (vessel, pipework and instrumentation) by use of collet connectors. Two connectors will mate by vertical entry to complete the inlet and export flow paths and the PSM can be pulled and re-run with the techniques and equipment used for running and pulling subsea Xmas trees.

For either version the non-recoverable base pipework includes a bypass loop with isolation valves to allow continued production without metering if the PSM is removed for any significant period. However the more likely policy would be to remove the PSM only in the event of a serious technical problem (or to clear an excessive sand build up) and to replace the unit immediately with a spare while the guidewires are in place and the diver or ROV is available.

For such operations the diver task must be minimised due to the need to work between tides in many typical gas fields locations. The actions will be to:

- Close off manual isolation valves at the PSM, (assuming production is already shut down).

- Disconnect the two flange connections at the inlet and outlet to the base pipework.

- Disconnect the umbilical jumper to the PSM control module by removal of one multi-stab flying lead connector.

- Attach a lift line and guidewires and monitor the lift.

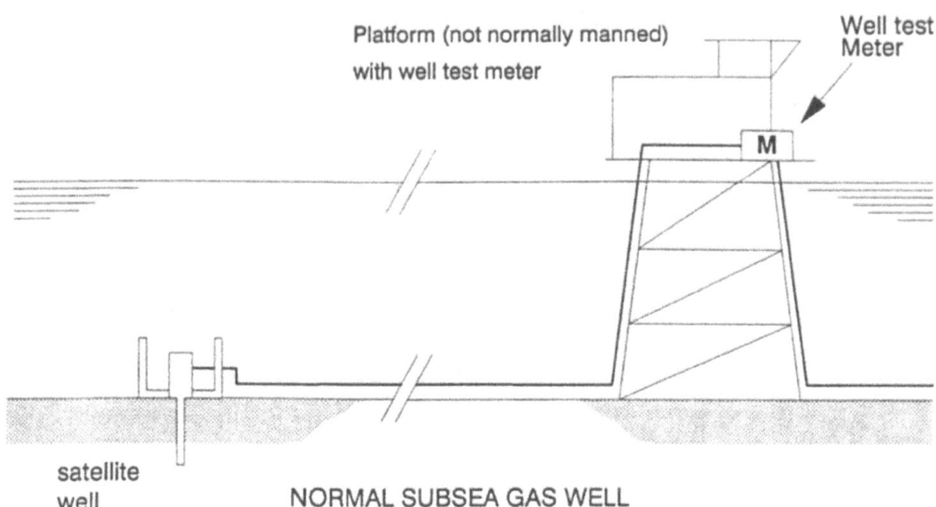

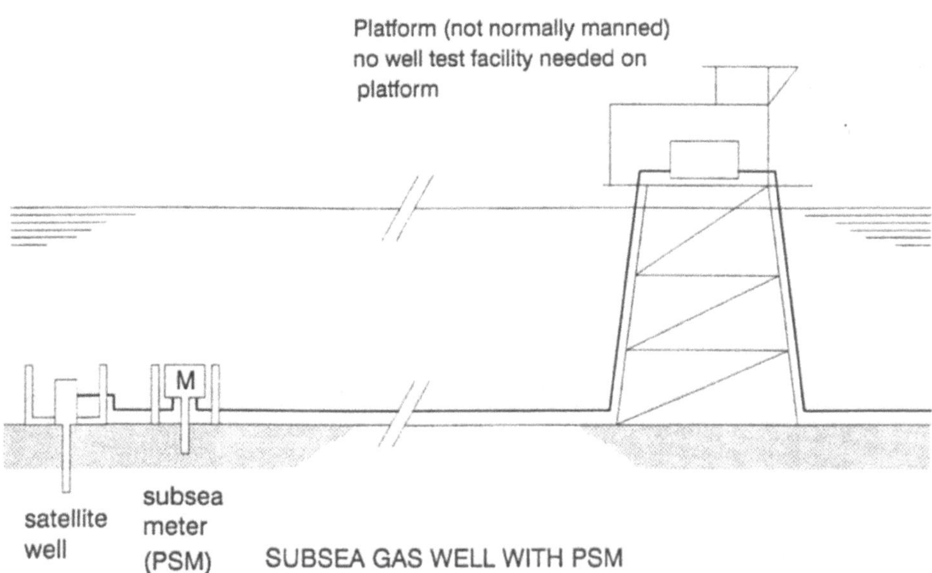

FIG 1 PHASE SPLITTER & METERING UNIT (PSM) SIMPLEST APPLICATION - SINGLE SUBSEA STEP-OUT WELL

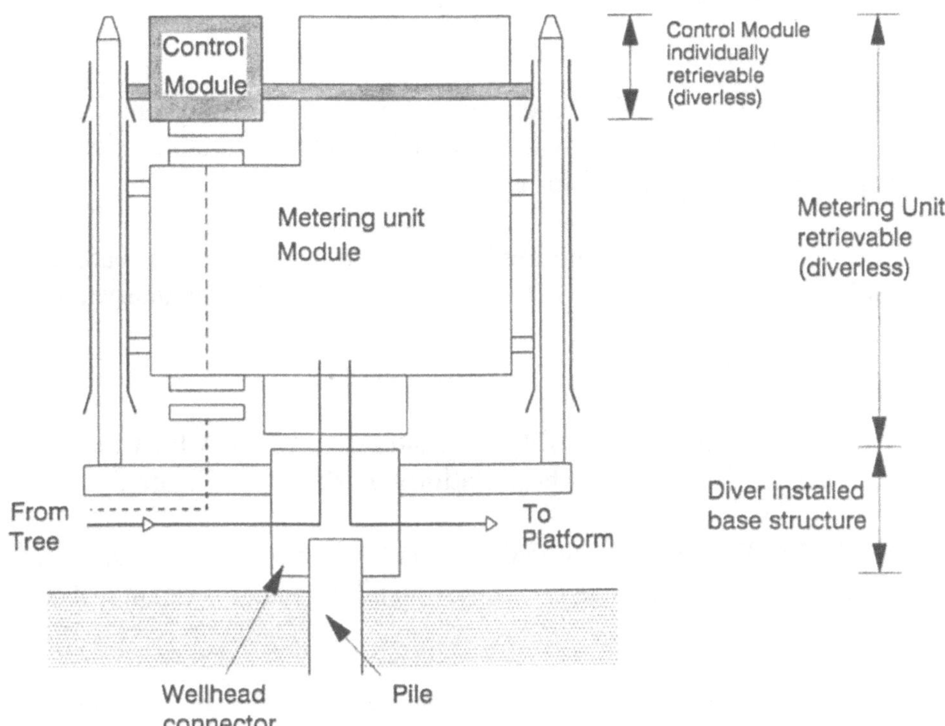

FIG 2 PHASE SPLITTER & METERING UNIT BASIC CONFIGURATION

- Complete the reverse procedure to install a replacement unit (probably at the next slack water period).

For the enhanced version the flange connect/disconnect actions are deleted. Diver time will be substantially reduced and could be eliminated since the other tasks can be engineered for an ROV.

While diver-friendly design and minimum cost replacement is a feature of the PSM, the development aim is to avoid any need for subsea intervention.

APPLICATIONS

The main application of the PSM is the subsea gas field which involves several wells all tied back to a host platform as shown by Figure 3.

If individual well testing is required each well will need separate test lines to the platform, or the alternative of a test manifold and a common test line as shown.

The test line is a major additional cost and the simplest way to avoid it is to make no provision for individual well test but to test "by difference". This means shutting down a well and deducing its production rate from the resulting decline in overall production.

While certainly a low cost solution this practice has operational disadvantages which can include loss of production and possible restart problems for some wells. It may also give inaccurate results due to any change in flowing conditions (eg back pressure) which can modify the production rate from the other wells.

Better data can be obtained by a positive metering of the production from each well. By providing a test header leading to a PSM each well can be metered but the test line cost can be eliminated.

The economic advantage will depend on the trade-off between the cost of the PSM unit and the pipeline, riser and topside equipment that it replaces. It will therefore depend on the number of wells and the distance to the platform.

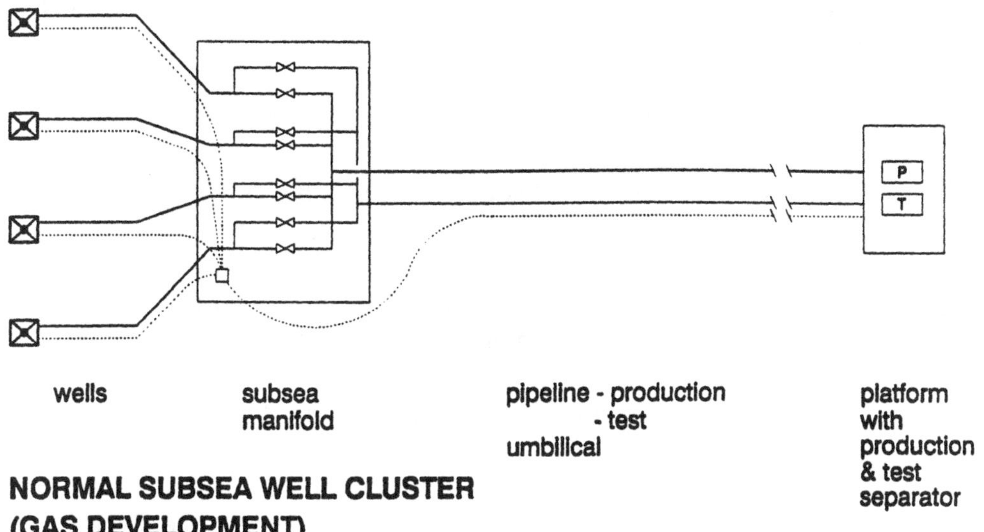

NORMAL SUBSEA WELL CLUSTER (GAS DEVELOPMENT)

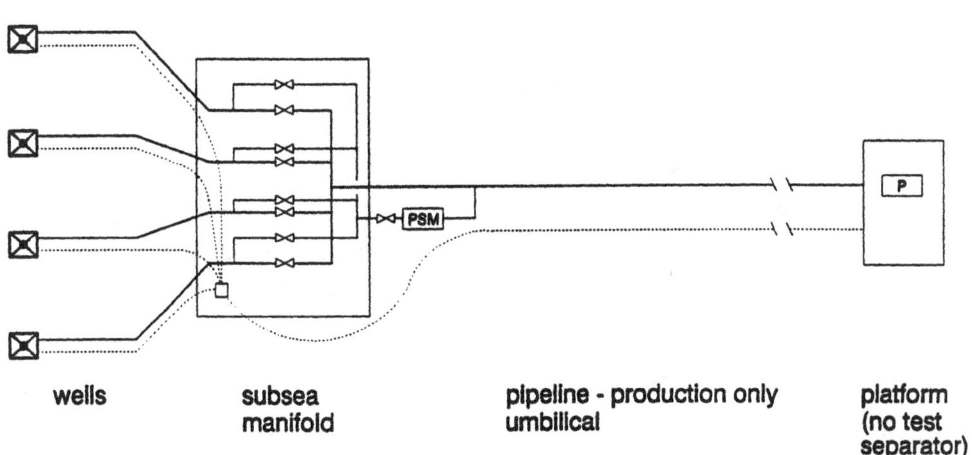

SUBSEA WELL CLUSTER WITH PSM

FIG 3 PSM APPLICATION

The capital cost benefits are supplemented by the operational benefits which include:

- Continuous on-line measurement of any well production.
- More accurate metering than a test line may provide (due to the elimination of test line back pressure effects).
- The opportunity to keep the liquids separate, which may reduce the consumption of hydrate inhibitor in the gas line.
- A reduction in platform topside weight, space and manning requirements.
- A structural base with provision for pigging later in the field life.

If the wells are in fiscally separate groups eg: separate fields, separate tax status or separate Operators a PSM can provide allocation metering. This can avoid the cost of providing separate pipelines to the platform, while at the platform it will avoid separate riser, J-tube and topside reception and metering trains.

As service experience is gained the PSM may be developed beyond the well test and allocation metering applications to provide full fiscal metering at the wellhead. This could widen the choice of reception platforms and trunk line tie-in options and may avoid the need for some platforms altogether.

TECHNICAL DESCRIPTION

Basic Principle

The process and instrument logic is shown in figure 4. Wellfluids enter the vertical separation vessel via an inlet choke. Gas exits from the top to enter the metering loop while liquid is collected in the bottom and drained out, continuously or intermittently, into a separate and much smaller metering loop before being reinjected into the export pipeline.

To achieve the reinjection of liquids without a mechanical pump the vessel pressure is maintained higher than the export pipeline entry pressure. For this purpose a differential pressure control valve (PCV) is included in the gas vent line. It is controlled by the programmable logic controller (PLC) in response to a differential pressure (dp) reading.

Gas mass flow measurement

Gas mass flow measurement is continuously provided by an ultrasonic flowmeter in the primary measurement run, with associated real time readings of gas pressure (P) and temperature (T). The flowmeter gives a velocity (v) dependent signal, while the pressure and temperature enable gas density to be derived given the molecular weight and the universal gas constant. Molecular weight is obtained from sample analysis at the platform.

Molecular weight is the only variable that is not a real time measurement but this is not expected to vary significantly from hour to hour or day to day. However to maximise the accuracy in measurement (and provide for future fiscal metering requirements) the PSM incorporates a sample collection facility.

The available methods of diver or ROV sampling may be used for the first operational units. More frequent samples could be transported via an umbilical hose to the platform if such a system can be developed. To develop it will require attention to material compatibility, purge systems and safety requirements. The benefit will be a regular sampling facility with the potential to meet fiscal metering requirements and without subsea intervention costs.

A periodic self checking facility is incorporated to remote verify or reset the calibration of the ultrasonic flowmeter. This takes the form of a parallel loop with a venturi or orifice plate.

At defined intervals the gas flow is remotely and automatically diverted through this secondary or check-calibration loop. The time averaged reading of velocity is compared with similar readings of the primary meter before, after and during the flow diversion cycle. Any consistent variation will be logged and used to reset the primary meter.

This arrangement utilises the advantages and avoids the disadvantages of both measurement systems. The ultrasonic flowmeter is non-intrusive which means it is not normally subject to wear and will not introduce a pressure drop. The venturi or orifice plate is a more proven device accepted as an industry standard but subject to wear in continuous service. By using it to check the ultrasonic meter the erosion effects are limited to short duration periods.

A venturi is technically preferable for a subsea unit but an orifice plate is a more established industry standard for topsides metering, accepted by Operators, gas customers, or the DTI. The design and selection of this reference meter depends on the constraints it imposes on the pipework layout and the frequency and ease of replacing the measuring section.

To maintain accuracy and reliability on all P, T and v measurements, triplicated sensors and channels with two out of three voting at the PLC is included as standard.

Liquid Mass Flow Measurement

Liquid mass flow measurement is complicated by the need to identify and measure both hydrocarbons and produced water. However the small quantities involved (in fiscal terms) make an acceptable standard of accuracy easier to obtain. A Coriolis meter is specified, supplemented by direct volumetric measurements of liquid hold up in the vessel.

By measurement of the changes in total liquid level and the oil/water interface level in the vessel the proportions of oil and water can be approximately verified. Level monitoring will be of the nucleonic type utilising a low radiation source inside the vessel and sensors located on the outside. This proven technique avoids any moving parts or vessel penetrations. Triplicated channels and two out of three voting is again a feature.

A sample take-off will be provided for diver or ROV operation. Locating the liquid and gas sampling points adjacent to each other can allow an ROV to dock a special tool and take both samples simultaneously. An umbilical hose route would be preferable as noted for the gas sample. With development a common hose may handle batch samples of gas and liquid.

Vessel Protection System

The need for the vessel to withstand full shut-in well pressure will result in thick walled and expensive vessels and can become impractical for abnormal well pressures. The preferred solution is a high integrity protection system (HIPS) which will close one or more upstream isolation valves to prevent the vessel being exposed to such internal pressures. A HIPS system is accepted practice on topsides plant and while a subsea HIPS may appear to add complexity it is the better development in the longer term. Once it is proven and accepted subsea the instrumentation cost will be modest compared with the savings in high rated vessel and pipework costs. These savings could also apply to the pipeline, risers, subsea pig launchers and other downstream elements.

A further safety advantage is the provision of an independently controlled barrier downstream of the production manifold and in addition to the platform isolation valve.

Pipework and Vessel

The vertical separation vessel and pipework may be rated to withstand full well pressure for the prototype unit, although later developments would incorporate a HIPS and allow a lower pressure rating as discussed above. Pipework is designed and developed to provide the required uniformity of flow at the measuring sections. This requirement dictates the extent of the straight lengths upstream and downstream of the sections and the need to avoid swirl. The pipework design is made more demanding by an operational need to limit the vertical height of the unit to that of an associated Xmas tree. This constraint will permit a common design of protection frame to serve both units, and minimise the structure cost which is height dependant.

Support Structure

The support structure comprises an industry standard 4-post guide base installed on a 30 inch stub pile by use of conventional subsea wellhead components. The PSM is supported like a simple Xmas tree, so actual components designed for well completions could be adapted for this purpose. Pressure tight sealing is not required, only structural support, so in principle the use of ex-wellhead components that cannot pass a pressure test may be considered (if available).

Control Module

The control module comprises upper and lower compartments. The upper compartment is a pressure resistant shell containing the PLC in a nitrogen atmosphere at 1.5 bar.a. The lower compartment is an oil filled pressure balanced enclosure containing servo valves, pressure transducers, supporting instrumentation and pipework. Electrical communication between the PLC and these items is via a fully water blocked connector in the dividing bulkhead. Communication with the adjacent tree or subsea manifold is by an umbilical jumper which terminates in either a set of diver connectable tails or an ROV matable multistab connector. Connections consist of:

- hydraulic supply and return
- nitrogen supply
- chemical injection
- instrument power supplies
- multiplex signal data highway
- sampling hose (gas and liquid)

Sampling hoses may be included in the umbilical even if the direct sampling system is not initially developed. The inclusion of the hoses would provide an option for a later retrofit if such a system is needed and developed in later years. If not needed for sampling the hoses will be retained as spare cores which are good practice generally.

The control module with its support frame is run on guidewires and located on two of the four guideposts. In service it can be recovered and replaced by ROV intervention without divers and without pulling the complete PSM.

Communication,power and service channels to the platform will be via the main umbilical.

Pigging

If pigging of the infield pipeline is required, provision for a subsea pig launcher can be included. As the PSM is a separate structure on a firm foundation, it provides a convenient site to install a removable launcher for sphere, scraper pigs, or an intelligent inspection pig.

The provision can take the form of a flanged pipe end with a piggable route to the infield pipeline. This basic pipework would include a closed isolation valve and a blind flange in place. If pigging is required an upward curving transition spool with a second manual valve will be added to provide double isolation. This would be diver installed and left in place. The pig launcher would be a recoverable unit lowered down guidewires to mate by vertical connection with the upward facing flange of the transition spool. The vertical barrel of the pig launcher would be exposed to damage when connected. It is therefore a temporary attachment only, being deployed down the guidewires in the loaded condition, connected by the diver and recovered after the pig is launched and the isolated valves are closed.

Both the transition spool and temporary pig launcher would be supplied as operational items and are not included in the capital cost of the PSM, which only includes the blind flange and one manual isolation valve.

RELATED DEVELOPMENTS

The PSM is derived from the subsea separation pilot unit tested by BOET on the "Argyll" field in 1988-90 in co-operation with Hamilton Brothers Oil and Gas (UK) Ltd. Many of the key elements described in this paper were demonstrated successfully by the 60 ton trial unit including Nucleonic level sensing, closed loop pressure and level control, the subsea PLC and use of triplicated key systems with 2 out of 3 voting. This trials unit is shown by Fig 5.

Two of the three vessels on the pilot unit were rated at less than full shut-in well pressure and were protected by a combination of pressure and level

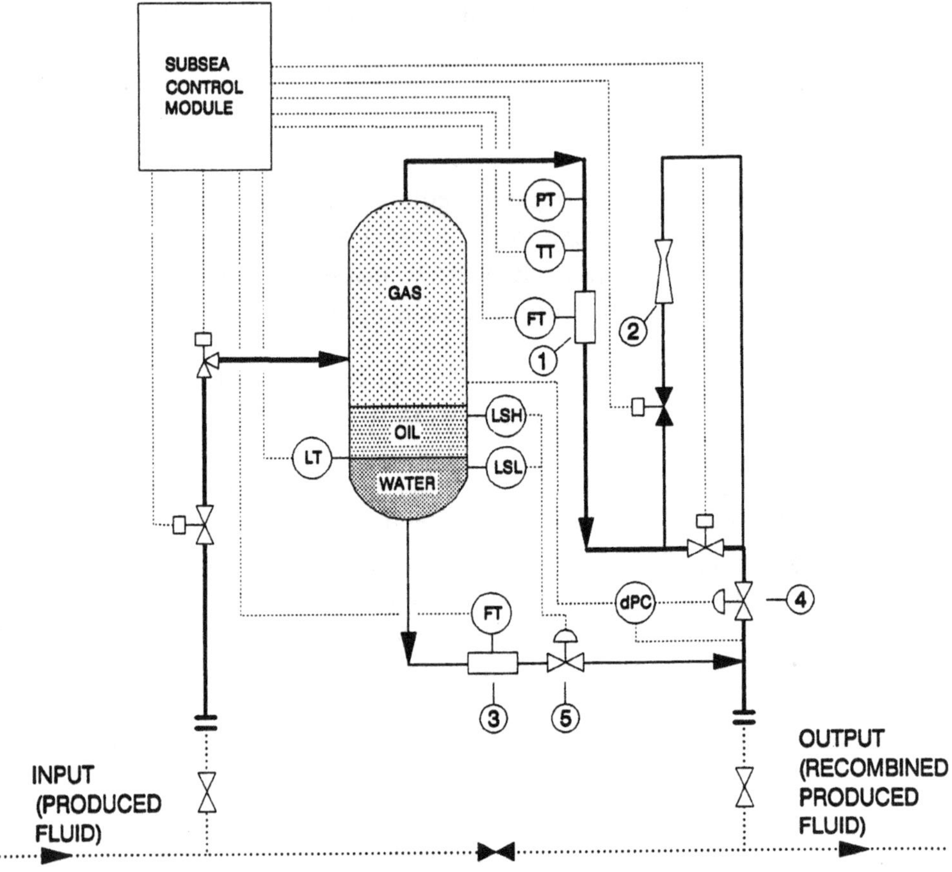

KEY

1. Ultrasonic Flow Meter (Gas)
2. Venturi (or Orifice Plate) Reference Flow Meter (Gas)
3. Coriolis Flow Meter (Liquid)
4. Pressure Control Valve
5. Liquid Level Control Valve

FIG 4 PROCESS AND INSTRUMENT SCHEMATIC (DIVER INSTALLED)

control. This fully certified unit therefore demonstrated the principle of a subsea HIPS design.

Related development include the Texaco subsea metering system (SMS) and the CBJ Volumetric flowmeter.

The Texaco SMS is also a separation and metering system but uses inclined tubular sections instead of a vessel. Gas velocity is measured by a vortex meter and liquid velocity by a proprietary Texaco constriction meter. Liquid density is measured by a gamma ray densitometer and oil/water ratio by a microwave technique. A 45 tonne prototype is being evaluated on the "Tartan" platform.

FIG 5 BOET SUBSEA SEPARATOR TRIAL 1988-89

The CBJ meter is non intrusive and uses gamma ray densitometer and capacitance sensor techniques. It incorporates a fraction meter developed with support from BP Norway and Saga Petroleum and tested on a BP land based field.

As hydrocarbon reserves dwindle in the UKCS new technology will be needed to maintain viable commercial exploitation. Subsea metering can be a key area for such technology.

REFERENCES

Edwards WG, Winter CL., (British Offshore Engineering Technology) UK Patent Application. "Subsea Separator, Storage and Pumping Unit and its Associated Control". (1990)

Dean TL, Dowty Dr. EL; Jiskoot M.A. (Texaco Ltd). "The Texaco Subsea Three Phase Metering System". (1990)

Dykesteen E (Chr. Michelsen Institute) "Multiphase Metering". (1991)

Eyre GP; Songhurst BW (British Offshore Engineering Technology) "Experience gained from the First UK Subsea Separator". (1990)

Flexible Pipe Materials for Aggressive Hydrocarbon Applications

R.T. Hill and J.C. Measamer
Wellstream Corporation
1700 C Avenue
Panama City, FL 32401
U.S.A.

Abstract

The increasing development of marginal offshore hydrocarbon deposits has resulted in production of increasingly corrosive fluids. This has increased the need for piplines capable of operating at elavated temperatures in the presence of high concentrations of H2S and CO2 gases. Conventional pipelines require the use of stainless steels or corrosion resistant alloys which drives up the cost of materials and fabrication/installation of the system. The use of flexible pipe systems is assuming an important role in these applications where stainless steel and CRA carcass materials may be economically utilized.

Volume 30: Subsea International '93, 109–123.

INTRODUCTION

This paper addresses the use of unbonded flexible pipe (figure 1) for aggressive environmental conditions. These conditions limit the use of rigid pipe to stainless or corrosion resistant alloys thus driving up the price of the pipe as well as expenditure for expensive corrosion inhibition systems (Hill, 1989, OTC 6114). In these types of environments flexible pipe can provide a cost effective and environmentally sound solution to field development problems.

A great deal of research has been completed over the past several years by the Force Institute and various material suppliers leading to an increased understanding of corrosion mechanisms in flexible pipe systems. The results of these studies are discussed later.

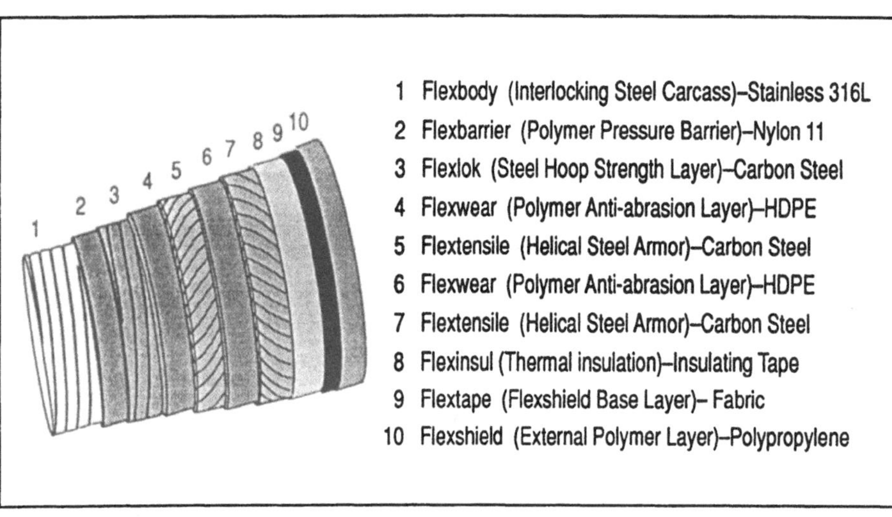

Figure 1-Typical Flexible Pipe Design

UNBONDED FLEXIBLE PIPE CONSTRUCTION AND MATERIALS

Each of the layers of unbonded flexible pipe is free to move relative to the others. This means that there is no composite action in bending or torsion. Each layer of the flexible pipe has a specific function to perform.

Carcass Layer

The primary purpose of the carcass layer is to prevent collapse of the flexible pipe. The collapse mechanism may occur due to either external hydrostatic pressure or the squeezing force of the helical armor wire layers under the influence of axial tension. For severe environments the material specified for carcass is generally 304L, 316L, Duplex or AL6XN alloy, however full-scale testing performed at the Force Institute, Copenhagen has shown that carbon steels are viable for a much wider range of process fluid corrosivity than previously believed. Under extremely severe conditions the carcass layer may be omitted and its role performed by the plastic barrier layer in conjunction with the steel hoop stress layer (McCone, 1989).

The carcass consists of an interlocked helix of steel strip which is pre-formed into an S-section (figure 2). The resulting structure is dimensioned to ensure stability and resist the greatest applied external pressure loads. When dimensioning the carcass a safety factor of 1.5 x the design external pressure is used.

Figure 2-Carcass Cross Section

The critical pressure for collapse is calculated from the classical formula for Euler stability of a solid cylinder with allowances included for ellipticity of the final formed shape as shown in figure 3 (B. Chen et al, FPT 92). The yield stress is additionally modified to account for the ratio between the average thickness of steel in the layer and the total thickness of the layer, the effect of yield value on collapse is shown in figure 4 and the effects of ellipticity on collapse pressure are shown in figure 5. These techniques have been validated by numerous full-scale tests on flexible pipes ranging from 2" I.D. to 10" I.D.

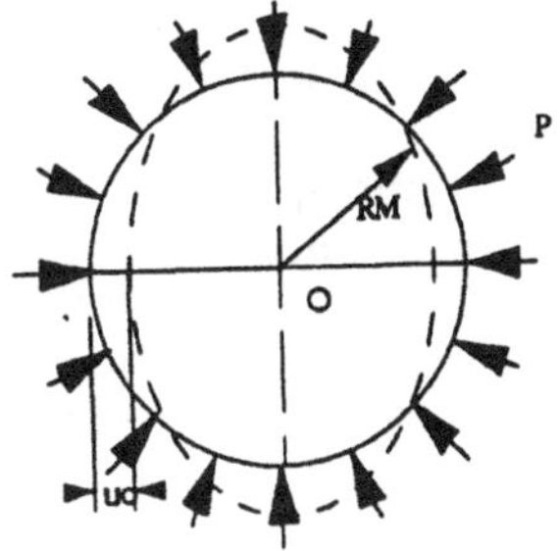

Figure 3-Elliptical Collapse Mode

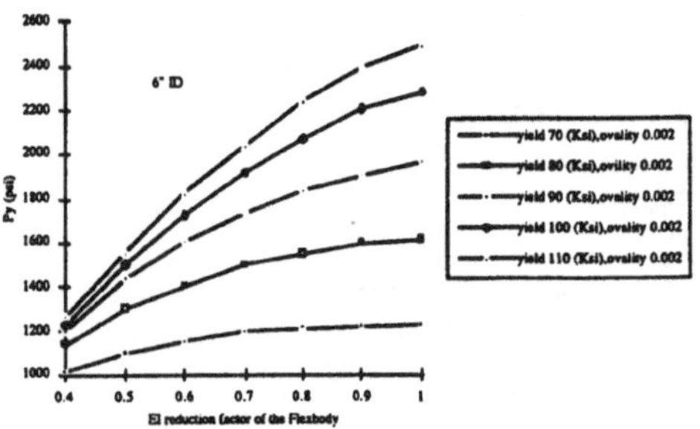

Figure 4-Influence of Material Yield

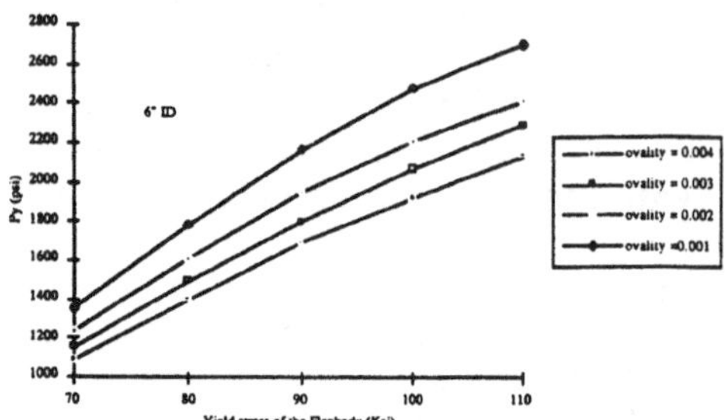

Figure 5-Influence of Ellipticity

Fluid Barrier Layer

The purpose of the plastic barrier layer is to contain the production fluids and gases. It is restrained against collapse by the carcass and against burst by the armor layers. Barrier layers may be specified as High Density Polyethelyne (HDPE), Nylon 11, Nylon 6/12 or Polyvinylidene flouride (PVDF) depending on the environment (Figure 6).

Material	UTS (MPa)	Yield (MPa)	Elastic Modulus (MPa)	Thermal Conductivity Upper Bounds (W/mK)	Application are for 10 yr life
HDPE	34	21	758	0,3	T-20/+ 60°C
Nylon 6/12	46.5		435	0.24	T-30/+ 80°C
Nylon11	55	29	345	0,3	T-40/+90°C
PVDF	41	29	758	0,16	T-40/+130°C

Figure 6-Plastics Properties

HDPE is generally specified for static applications where the temperature is less than 60 C for moderate pressure non-corrosive applications. For temperatures up to 80 C Nylon 6/12 may be specified, and Nylon11 is used for temperatures up to 100 C. One of several different PVDF polymers may be specified for temperatures up 135 C depending on other environmental conditions.

The thickness of the barrier layer must be specified so that the plastic will not yield due to internal pressure where it spans the gap between adjacent armor wires. For a given thickness of the plastic the critical gap width will decrease when a circumferential armor layer is specified, and the required thickness is limited to the thickness required to ensure that age-hardening cracks will not penetrate the wall during the lifetime of the pipe.

For extreme conditions when a carcass layer is not specified the barrier may nonetheless be stable under the influence of net hydrostatic pressure provided that a circumferential armor layer has been specified to restrain it against Euler collapse due to "rosetting".

Circumferential Pressure Armor Layer

The purpose of the circumferential armor layer is to support the barrier layer against burst under the influence of internal pressure. In low pressure applications this purpose may be served by the helical armor layers, but in high pressure applications the need to ensure that the width of unsupported plastic in the armor wire gaps dictates the use of an interlocked circumferential armor layer. Figure 7 shows the interlocking profile of the flexlok layer.

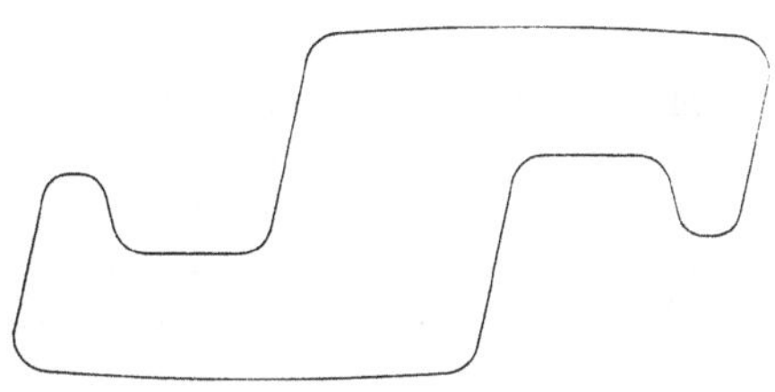

Figure 7-Typical Circumferential Pressure Layer

Since there is always a double layer of helical armor the circumferential armor has only to contribute a portion of the total burst resistance of the pipe. For ease of manufacturing the circumferential armor layer is of low-carbon steel with a typical yield stress of 758 MPa. Figure 8 shows a free body and force equilibrium diagram of stresses resulting from the application of internal pressure. This diagram includes the forces resulting from free movement of the end fittings. In this case the axial force on the end fittings is transferred directly to the helical armor layer resulting in a squeeze pressure being applied to the circumferential armor(B. Chen et al, FPT 92).

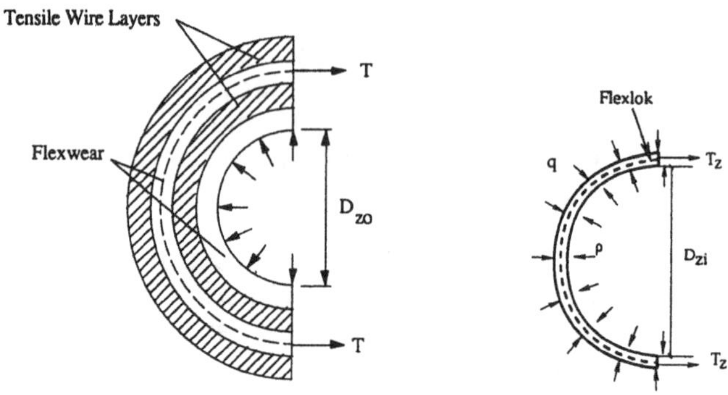

Figure 8-Free Body Force Diagram

When the burst pressure requirement is high a second layer of of circumferential armor may be used to supplement the interlocked circumferential armor layer. Since this layer is not in direct contact with the barrier layer, it does not have to be interlocked and can therefore be made of a simple rectangular shaped wire with mechanical and chemical properties similar to the circumferential armor layer.

Helical Armor Layer

The helical armor layers perform two functions. The first is to support the plastic barrier layer against burst when a circumferential armor layer is not specified. The second function is to provide the pipe with axial strength to enable it to withstand the end-cap force effects of internal pressure and to resist other tensile loads applied to the pipe (McCone 1989).

There are always two contra-wound layers of helical armor wire in order to balance the torsion with the lay angle of the two layers are adjusted to account for the difference in centerline radius. The wire is pre-formed in torsion and bending so as to minimize the residual elastic stresses and prevent the wires from springing back (birdcaging) when the pipe is cut.

In the absence of a circumferential armor layer the helical armor must withstand the hoop and end-cap effects of internal pressure alone. Under these circumstances the "balanced" lay angle under which there is no tendency for the armor helix to change shape under load is 54.7 (Colquhoun, Hill and Nielsen, 1990).

When a circumferential armor layer is present the relationship changes and the "balanced" angle depends on the proportions of steel in the circumferential armor and helical layers, and the relative strengths of the wires. The optimum lay angle is then typically 35 to 45 . The failure tension of the pipe may be governed by the collapse of the carcass or by yield of the helical armor wires depending on the characteristics of these two layers.

The helical armor wires typically have cross-sections ranging from 5 x 2mm to 10 x 4mm. While the smaller sections are easier to apply pipes of larger diameter require greater cross-sections in order to provide the required hoop and tension strength. The wires may be of either low carbon (758 MPa) or high carbon (1550 MPa) steel based on the process conditions. Low carbon steels are indicated when process fluids are sour since hydrogen sulfide and water vapour will readily diffuse into the armor annulus.

Tape Layer

A thin tape layer is used around the outer helical armor layer to provide a means of preventing penetration of the final extruded shield layer into the armor interstices and ensure that the shield does not bond with the helical armor.

Outer Shield Layer

The primary purpose of the outer shield layer is to protect the helical armor from mechanical damage and exposure to seawater. The shield layer may also contain "burst disks". These are small circular areas milled into the shield to provide reduced wall thickness. These areas are designed to "pop" in the event of a buildup of diffused gases in the armor anulus. Since the sea water intrusion will be rapidly depleted of oxygen and cannot circulate, when combined with suitable cathodic protection there is no danger of severe corrosion occurring.

The primary material used for shield layers in static applications is HDPE due to its low cost and and resistance to attack by fresh and salt water. HDPE may be specified with either black or orange color when resistance to UV is required.

Nylon 11 has also been shown to be impervious to attack by fresh and salt water (Coflexip and IFP, OTC 5231). Nylon 11 is more flexible and tougher than HDPE and is normally specified for dynamic riser applications, however its 5:1 cost ratio relative to HDPE makes it un-economic in static applications. Nylon 6/12 material has now been introduced to the flexible pipe market and provides most of the benefits of Nylon 11 at a 20% cost savings.

Insulating Layer

While unbonded flexible pipe has high inherent insulating properties the possibility of wax dropout or hydrate formation will sometimes require additional insulation.

Insulation materials must not only slow the heat loss but must be able to withstand hydrostatic pressure and exposure to sea water if the outer shield layer is damaged. Additionally the material must be able to withstand the extrusion temperatures of the shield layer without degrading or out-gassing. Wellstream Corporation has manufactured a specialized test chamber for testing of insulating material. Figure 9 shows a typical system test configuration, this specimen is then placed into a large chamber for full scale testing.

Insulating layers are generally constructed of either PVC, for low temperature applications, or syntactic foam. This layer consists of an extruded or cast interlocking profiled strip which is wound over the final helical armor layer.

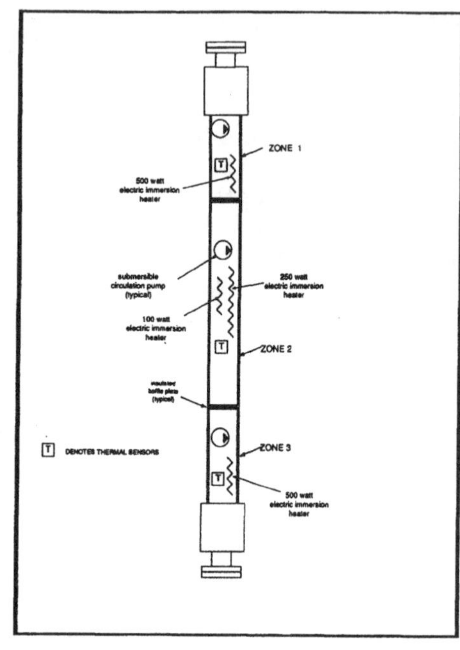

Figure 9-Full Scale Thermal Testing

End Termination

The primary function of the end terminations (figure 10) are to prevent loss of fluid from internal pressure and to prevent the entrance of sea water into the annulus of the pipe. End fittings are normally provided with pressure relief valves to vent any permeated gases from reaching a pressure which would damage the external shield layer.

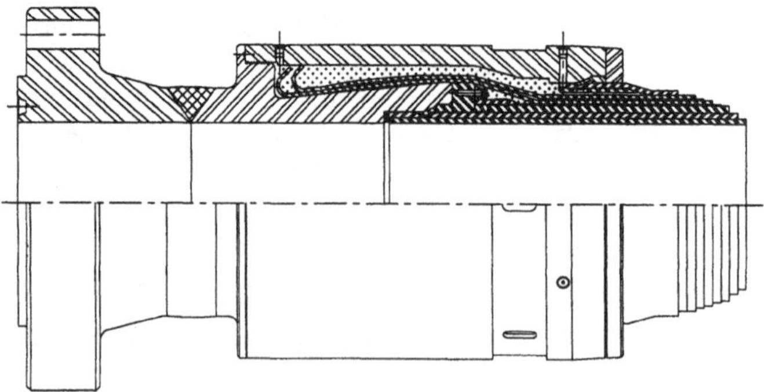

Figure 10-Typical End Termination

End terminations are normally manufactured of AISA 4130 steels, for severe environments a cladding of Inconel 625 may be added to the wetted surfaces, when conditions are most severe Duplex steels may be specified. Extensive testing has been performed to validate both the end fitting design and its connection to the pipe. As an example a recent 10" I.D. flexible required over 740,000 lbs of tension to damage the helical armor wires, this test resulted in no damage to the end termination.

MANUFACTURE AND QUALITY SYSTEMS ANALYSIS

Flexible pipes operating in the production and transport of petrochemicals must meet the highest standards of quality. A quality system should be in place which meets the requirements of ISO 9001/ANSI Q91 or an equivalent standard. The use of these standards will ensure that the manufacturer is capable of designing and manufacturing a suitable product which will withstand the design environment.

Manufacturing methods for metallic layers include forming carcass materials from a flat strip into a s-shaped interlocked layer, and applying pre-forming to circumferential and helical armor wires to ensure that they lay correctly on the pipe. Extrusion of plastics layers are closely controlled to ensure that the desired properties of the material are retained. Extensive testing on as-extruded material has been performed to verify material properties.

All flexible pipes are factory proof tested to 1.5 times rated working pressure for a period of 24 hours. Pressure testing of flexible pipes requires a conditioning period where the pipe pressure is raised and lowered systematically in order to reach an equilibrium point. This ensures that the pipe pressure does not drop below test pressure due to expansion of the pipe.

FULL SCALE TESTING

Dynamic Applications

Full scale testing of dynamic risers are currently being performed at the SIPM facility in The Hague (figure 11). This test will provide accurate long term data on the ability of riser designs to perform in severe motion environments over an extended period of time. This testing is designed to simulate 20 years service.

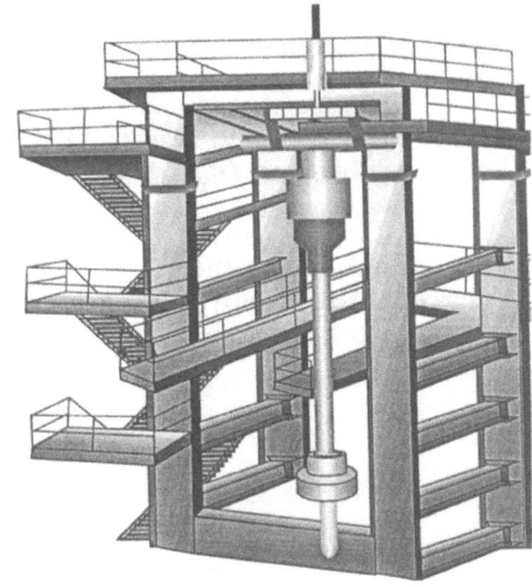

Figure 11-Dynamic Test Unit

Corrosion Testing

Extensive corrosion testing has been performed by The Force Institute to measure the effects of sweet and sour fluids on the corrosion of both carbon and stainless steel carcass materials. Additional testing in highly corrosive environments was conducted on various high strength helical armor wires. The Force Institute has developed a test rig (figure 12) which enables testing of flexible pipes under varying partial pressures of H2S and CO2.

Full scale testing of carbon steel carcass sections has shown that this material is able to withstand more severe environments than previously anticipated. A total of five tests with varying compositions (figure 13) were performed continuously for a period of 14 days each. The stainless steel samples were unaffected by any of the five environments.

In test 1 the carbon steel carcass exhibited a black corrosion film and was lightly etched on the surface, the corrosion film was analysed using X-ray diffraction and shown to be ferrous sulfide and ferrous carbonates. Weight loss of the eight test coupons were 8 +- 0.15 mpy.

In test two, four and five the carbon steel exhibited no signs of corrosion.

In test three the carcass appeared to be lightly etched and showed a weight loss of 19 +-0.15 mpy.

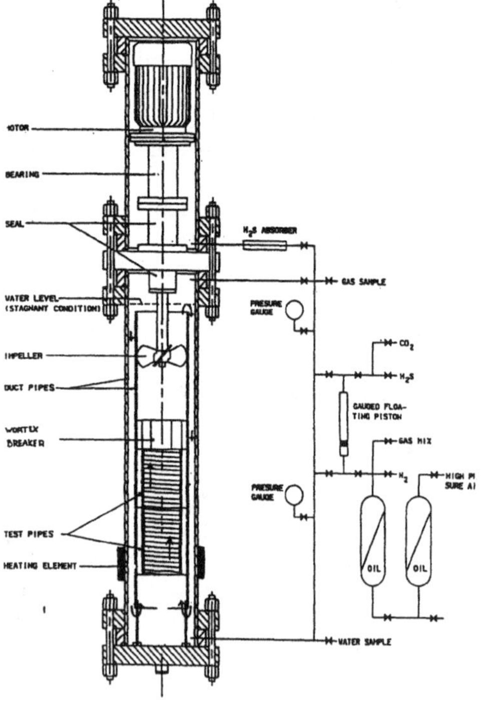

Figure 12-Corrosion Test Chamber

Test No.	Test designation	Test Brine	Solution Additions	Temperature, °F	Flow velocity, ft/sec.	Total pressure, Psi	Partial pressure, Psi H_2S	CO_2	N_2
1	Sour	4% NaCl	-	150	12	3100	100	500	2500
2	Sour + condensate	4% NaCl	10% Vol% condensate, C_7H_{16} - $C_{12}H_{26}$	150	12	3100	100	500	2500
3	Sweet	4% NaCl	-	150	12	3100	-	200	2900
4	Sweet	4% NaCl	COREXIT 3340 Water soluble inhibitor 50 ppm	150	12	3100	-	200	2900
5	Sweet, inhibi-	4% NaCl	COREXIT 7798 Oil soluble, water dispersable inhibitor 100 ppm	150	12	3100	-	200	2900

Figure 13-Corrosion Test Matrix

Wear Rate Testing

AEA Petroleum Services of the U.K. Atomic Energy Authority conducted "pigging trials" of flexible pipes using a 6" I.D. flexible in a 85 meter test loop using both gauging and cleaning pigs. The test loop used the flexible pipe at its minimum working bend radius and performed 2000 circuits to simulate 170 km of travel on the pig. A sophisticated Thin Layer Activation (TLA) technique was utilized to measure the material loss in the carcass of the flexible pipe (figure 14).

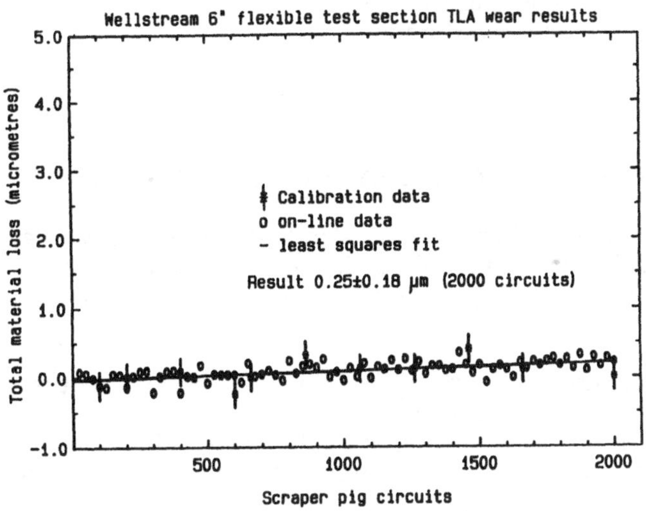

Figure 14-TLA Wear Results

This study showed negligible wear on both the pigs and the flexible pipe during the 170 km test.

ON-LINE MONITORING AND INSPECTION TECHNIQUES

Condition monitoring and inspection systems for flexible pipelines are currently being researched by several organizations. Presently there are three primary methods of inspection and monitoring that show promise (SINTEF, FPS 2000, 1992), these include test pipes with coupons, eddy current and ultrasonic imaging. Additional inspection methods which have been reviewed include TV holography, in-situ radiography and acoustic emission.

At present the use of test pipes with test coupons (figure 15) appears to be the most cost effective method of monitoring pipeline condition. The test pipe method requires installation of a curved section of pipe which can be shut off from the fluid flow for retrieval of test coupons. The coupons may include sections of carcass material as well as fluid barrier samples. This provides a means of monitoring both erosion and the long term effects of the conveyed fluid on the materials in use.

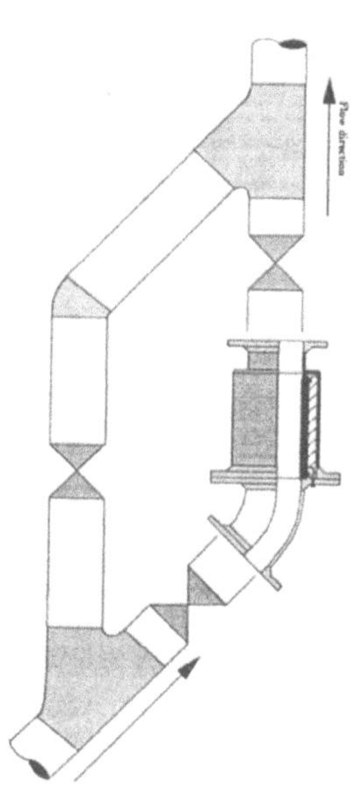

Figure 15-Test Pipe Monitoring

INSTALLATION AND MAINTENANCE CONSIDERATIONS

Flexible pipe may be laid using either standard reels or carousal systems when extended lengths are required. For installations in remote areas flexible pipes may be installed using powered A-frames mounted to available workboats.

In general terms flexible pipes may be laid at much lower cost then either reeled or conventional rigid pipe. Flexible flow-lines may often be laid without the need for pre-sweeping of the flowline path or use of support spans as the flexible will tend to conform to the seabed topography.

Flexible pipes are generally able to be laid at much higher speeds than either reeled or conventional rigid pipe, typically a vessel such as the NOS Norlift (figure 16) can install flexible pipe at a rate of approximately 10 km per day. This compares favourably with a typical reel lay vessel of 2 km per day. Studies conducted by Celant and Lazzari have shown that flexible pipes may be laid in 1/3 to 1/10 the time required for rigid pipes depending on the material selected.The higher rate enables the operator to take advantage of shorter weather windows which are not possible with other flowline systems.

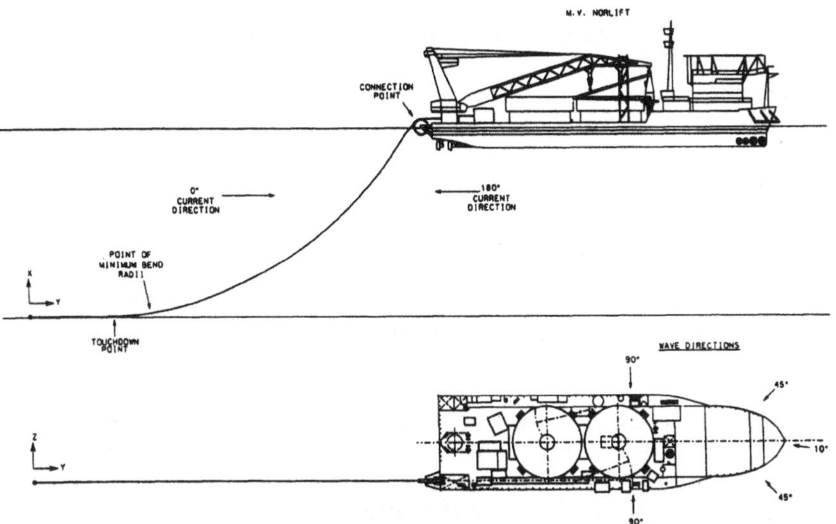

Figure 16-MSV Norlift Installation Vessel

COST ASSESSMENT AND COMPARISON OF FLEXIBLE PIPE SYSTEMS

Comparisons between towed bundles, reeled rigid pipe and flexible pipe show that flexible pipe provides a cost advantage for lengths ranging up to 10 km. when the installed total cost is used. In particular the use of flexible pipes in environments where upheaval buckling, mud slides or rugged topography may be encountered may show significant savings to the operator.

Flexible pipes are also cost effective for marginal field development and where flowlines and risers will be in temporary use. The ability to retrieve and redeploy flexible pipes is a major advantage under these circumstances. The decision on how to develop a particular field are influenced by many factors including:

- Length of flowline
- Field development plan (Floating production system or fixed platform)
- Topography
- Water depth
- Conveyed fluid temperature and content
- Presence of corrosive fluids and gases
- Availability of installation and support vessels
- Length of time required for installation
- Useful field life

Celant and Lazzari performed cost analysis of two types of flexible pipes (duplex S.S. and Inconel 825 carcass) along with carbon steel, duplex stainless steel and clad rigid pipes. The criteria used for comparison included acquisition cost, installation cost, design life, operating cost and replacement cost. Over a 25 year life flexible pipe provided the lowest life-cycle cost when used with a duplex stainless steel carcass for lengths of up to 10 km. A significant factor is the installation cost which showed the following factors:

	Laying Time Index	Laying Cost Index
Carbon Steel	1.00	1.00
Duplex S.S.	3.50	1.71
Clad C.S.	5.00	1.95
Flexible pipe	0.29	0.30

The studies by Celant and Lazzari are significant in that they developed cost structures which could be varied to provide a realistic cost comparison. Their study enabled use of varying discount rates, design life and pipeline length in order to accurately determine the life cycle cost of each type of pipeline system.

Studies conducted by Andrew Palmer & Associates in 1992 also show that flexible pipe is very cost competitive with rigid reeled pipe and towed bundles for lengths of up to 10 km. This study compared rigid pipe and flexible pipe in several different sizes and for varying production fluids, also compared were multi-bore towed bundles and flexible pipe.

For applications such as jumpers and spoolpieces flexible pipe is far more cost effective than rigid pipe, in these cases the need for exact orientation and measurement as well as the use of position alignment equipment can be eliminated.

References

Hill R.T., and Christensen, C.:"Full Scale Testing of Line Pipe Steels for Sour Service", New Orleans, 1989 (Corrosion 89).

Colquhoun, R.S., Hill, R.T., and Nielsen, R.:"Design and Materials Consideration for High Pressure Flexible Flowlines", Society for Underwater Technology, 1990.

Celant, M., and Lazzari, L.:Corrosion Economics in Material Selection for Sour Service Pipelines". Proceedings, 8th International Conference, BHRA, 1990.

Chen, B., Nielsen, R., and Colquhoun, R.S.:"Theoretical Models for Prediction of Burst and Collapse and their verification by Testing". Flexible Pipe Technology-International Seminar on Recent Research and Development, Norway, 1992.

Berge, S. ,"Inspection and Condition Monitoring", Flexible Pipe Technology-International Seminar on Recent Research and Development, Norway, 1992.

McCone, A.I., "Technical Notes on Unbonded Flexible Pipe Product Design Methods", San Francisco, 1989, Wellstream Corporation Technical Note Series.

Kalman, M.D., Alexander, W.L., Belcher, J.R., Loper, C.L., "Thermal Insulation for Flexible Pipe-Development and Testing", ASTM, Publication scheduled for ETCE 93, Houston.

AUTONOMOUS CONTROL SYSTEM (SPARCS)

FOR

LOW COST SUBSEA PRODUCTION SYSTEMS

BY

M THEOBALD

FSSL LIMITED, LONDON

SUMMARY

This paper describes a Subsea Powered Autonomous Remote Control System (SPARCS) which is designed to control subsea wells without the use of electro-hydraulic control umbilicals.

With oil exploration moving towards marginal fields, and fields connected to the existing infrastructure, SPARCS will provide a low cost solution to justify subsea developments.

The paper looks at each of the main system components and provides commercial justification for such a control system.

SPARCS is highly innovative in that it controls hydrocarbon wellheads without the use of control umbilicals. The systems electrical power subsea is produced by a Turbine Generator fitted to a water injection flowline or alternatively a Thermo-Electric Generator fitted to a production flowline. Hydraulic power for operating wellhead and downhole safety valves is produced from a subsea unit, with communication signals for valve control and sensor monitoring using acoustic telemetry with seawater as the transfer medium.

1.0 INTRODUCTION

The aim of SPARCS is to reduce the cost of future subsea hydrocarbon production developments by utilising innovative technology which does not require electro-hydraulic umbilicals from the platform to perform subsea wellhead control.

Over a period of 10 years it is estimated the SPARCS system would reduce overall costs by £72.5million and additionally produce hydrocarbon reserves of £87million per annum, which would not be economically feasible with existing technology. This is based on 5 single well developments per year at an average distance of 7.5km. The SPARCS system is shown in section 3.0 to be a commercially viable project.

As part of a joint industry programme, see Figure 1, Enterprise Oil, EE Caledonia and FSSL are completing a demonstration project to fully qualify the system.

EE Caledonia have made available the use of a water injection well no. 15/17-22Z in the Saltire North Sea Development to demonstrate the system over a period of 18 months.

The SPARCS system will comprise of two groups of components, the Surface Controller and the Subsea Control System. The Surface Controller forms the control station which provides the operator with the means of controlling and interrogating the subsea system. The Subsea Control System forms the outstation which provides all the control and data monitoring functions for the well.

Volume 30: Subsea International '93, 125–148.

Subsea power is produced with a Turbine Electric Generator fitted to a water injection well flowline or a Thermo-electric Generator fitted to a hydrocarbon production flowline depending on the type of well to be controlled.

A local Subsea Hydraulic Power Unit provides power to operate the full range of wellhead and downhole safety valves.

Acoustic telemetry is used to communicate with the platform up to a distance of 10Km, this distance is expected to increase up to 30Km with the development of other non-umbilical communication techniques.

The SPARCS system becomes economically more attractive when compared with long lengths of umbilicals in the field development, but breaks even on capital outlay with existing technology when the total umbilical distances would be approximately 2.3Km.

1.1 General Description - Existing Technology

A typical design for a subsea control system to allow production of hydrocarbons is shown in Figure 2.

This consists of a platform based operator console, wellhead Subsea Control Module (SCM), and electro-hydraulic umbilicals, together which control and monitor a range of wellhead hydraulic valves and production sensors (typically pressures and temperatures).

This type of control system, known as multiplexed electro-hydraulic, has become the predominant method of controlling subsea hydrocarbon production, due to its fast hydraulic response time, (permitting safe/reliable operation) and flexibility in respect of the types and configuration of production sensors. The SCM for this type of control system is dependant on the reliable operation of the interconnecting control umbilical to supply three indispensable functions:

1) Supply hydraulic control fluids, either water or mineral based, at pressures in the range of 200 to 1000 bar.
2) Supply of electrical power typically 220VAC 60Hz up to 200 watts for a single SCM.
3) Electrical two-way communications, typically Frequency Shift Keyed (FSK) to command valve operations and receive monitored sensor values.

The loss or degradation of any of these facilities could result in a field shutdown with associated loss of hydrocarbon production revenue.

As an overall system, the costs associated with umbilicals are a major proportion of the capital, installation and maintenance costs.

1.2 SPARCS Technology

A typical application of the SPARCS system is shown in figure 3. The significant design difference between SPARCS and existing systems is the total lack of any control umbilical between the platform and wellhead Subsea Control Module. This unique operation is obtained by suitably packaging a subsea electrical power generation unit and subsea hydraulic power unit (HPU) together with an acoustic telemetry system in order to perform two way communications using sea water as a transfer medium.

The research and development aspects of this system have already been completed, with evaluation reports available. The outstanding scope of work involves packaging the equipment to operate in a subsea environment, and a demonstration to evaluate and prove the overall integrated system in realistic operating conditions.

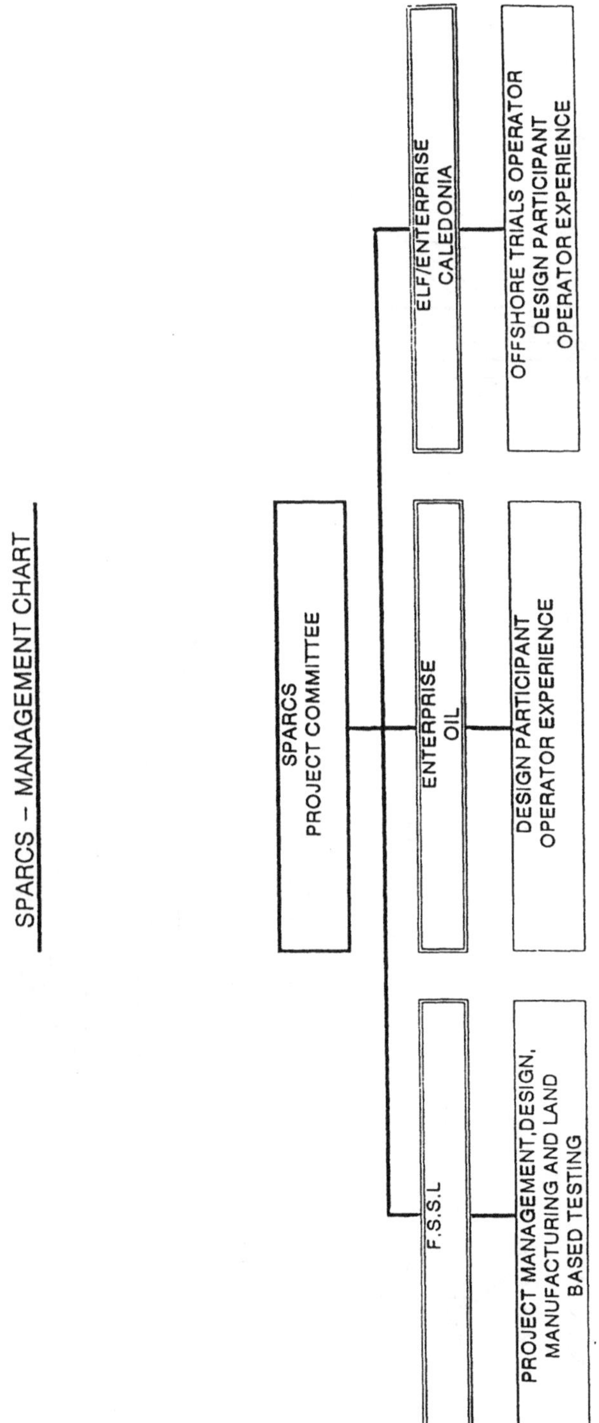

Figure 1 SPARCS - Joint Industry Programme (Management Chart)

SUBSEA HYDROCARBON PRODUCTION CONTROL SYSTEM - EXISTING TECHNOLOGY (SIMPLIFIED)
MULTIPLEXED ELECTRO-HYDRAULIC CONTROL
OPERATOR CONTROL CONSOLE
COMMUNICATIONS MODEM
ELECTRICAL POWER UNIT
HYDRAULIC POWER UNIT
PLATFORM FACILITY
TOPSIDE
SUBSEA
WELL FLOWLINE
MONITORING SENSOR (PRESSURE OR TEMPERATURE)
DEPTHS TO 250m
HYDRAULIC VALVE ACTUATOR
SUBSEA ELECTRONICS MODULE
ELECTRO-HYDRAULIC UMBILICAL LENGTHS UP TO 30Km
SOLENOID HYDRAULIC CONTROL VALVE
SUBSEA CONTROL MODULE
HYDRAULIC LINE
ELECTRICAL LINE
COMMUNICATION LINE

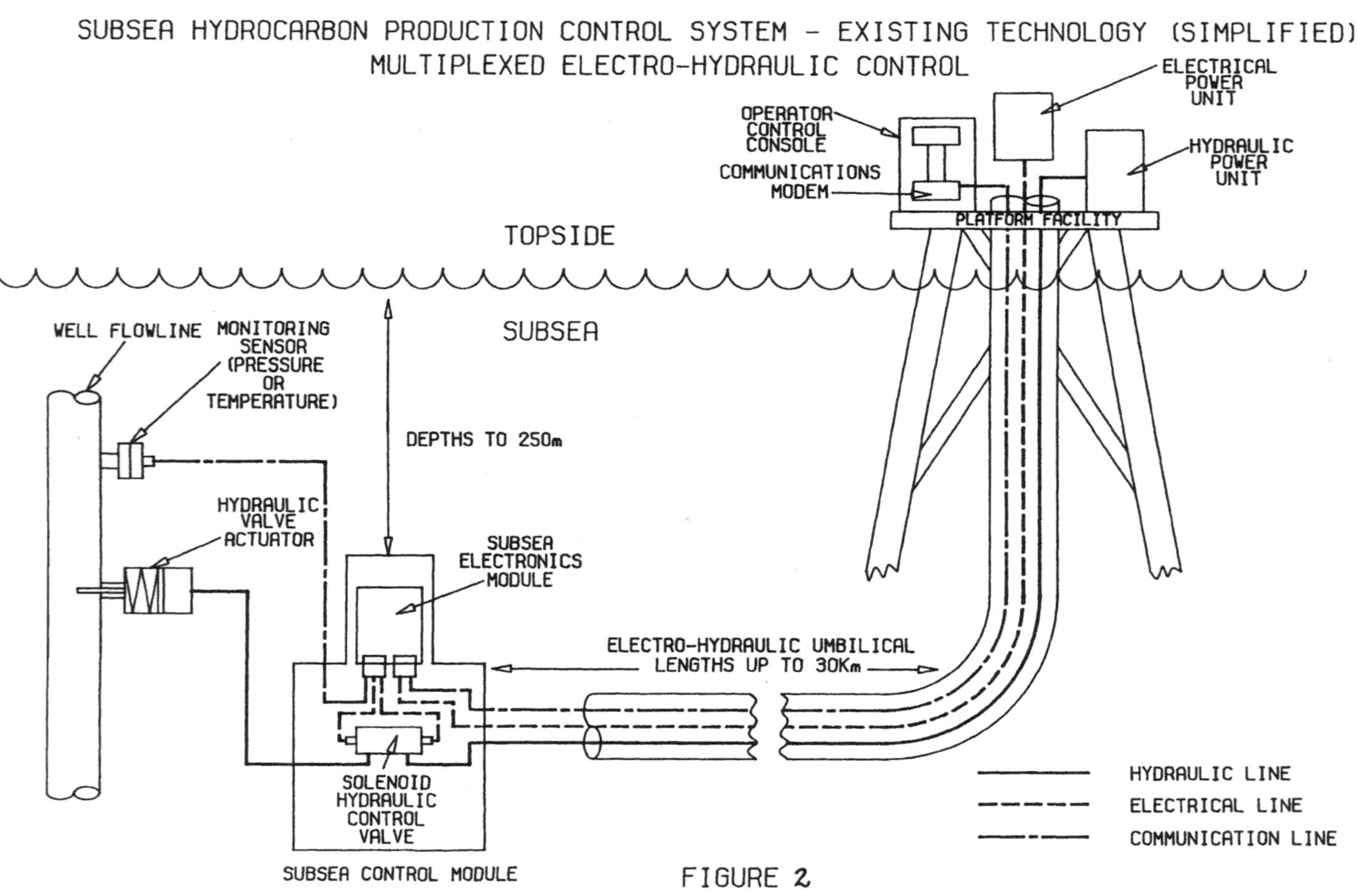

FIGURE 2

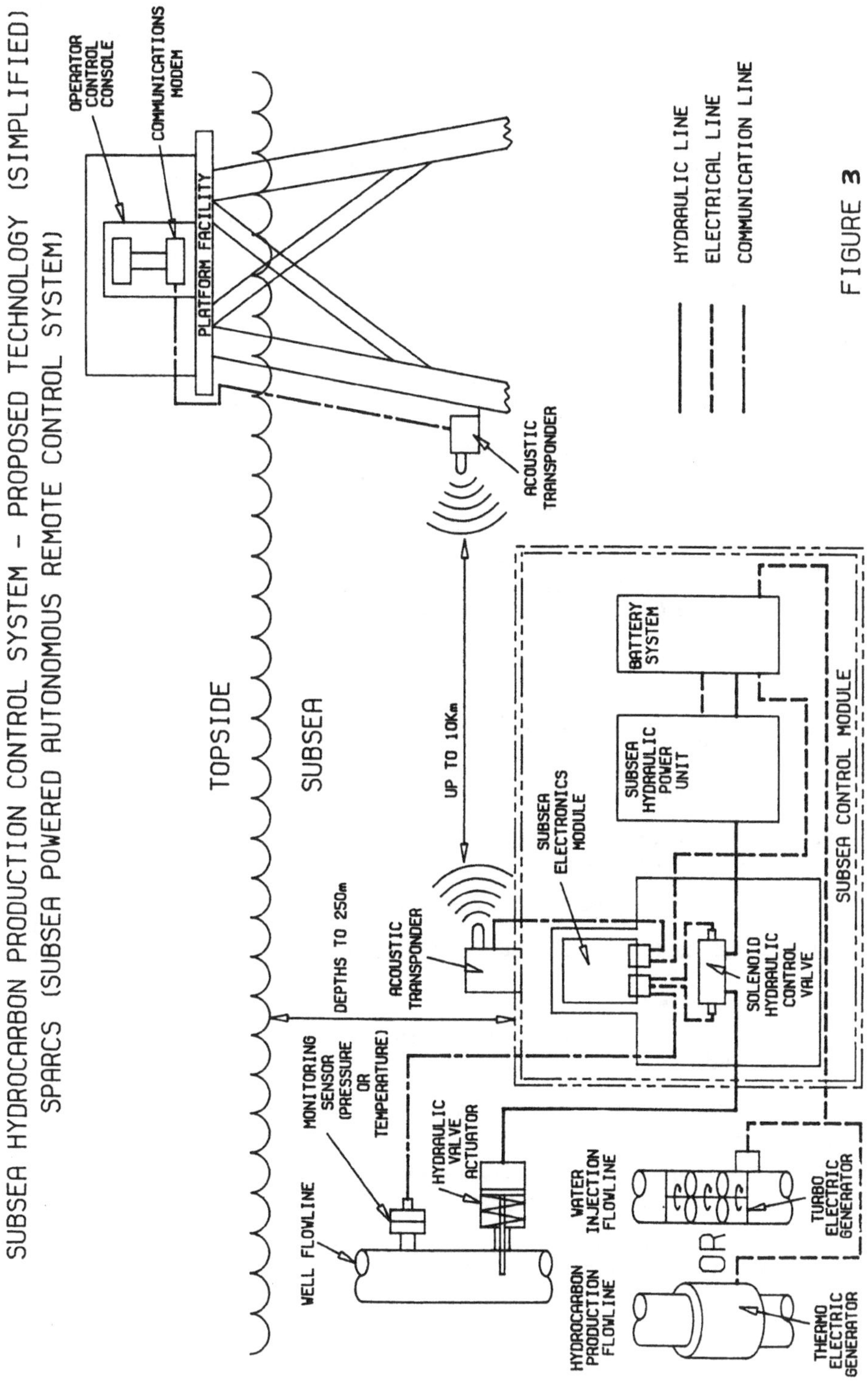

SUBSEA HYDROCARBON PRODUCTION CONTROL SYSTEM - PROPOSED TECHNOLOGY (SIMPLIFIED)
SPARCS (SUBSEA POWERED AUTONOMOUS REMOTE CONTROL SYSTEM)

OPERATOR CONTROL CONSOLE
COMMUNICATIONS MODEM
PLATFORM FACILITY
TOPSIDE
SUBSEA
ACOUSTIC TRANSPONDER
UP TO 10Km
DEPTHS TO 250m
ACOUSTIC TRANSPONDER
SUBSEA ELECTRONICS MODULE
BATTERY SYSTEM
SUBSEA HYDRAULIC POWER UNIT
SOLENOID HYDRAULIC CONTROL VALVE
SUBSEA CONTROL MODULE
MONITORING SENSOR (PRESSURE OR TEMPERATURE)
HYDRAULIC VALVE ACTUATOR
WELL FLOWLINE
WATER INJECTION FLOWLINE
TURBO ELECTRIC GENERATOR
OR
HYDROCARBON PRODUCTION FLOWLINE
THERMO ELECTRIC GENERATOR

HYDRAULIC LINE
ELECTRICAL LINE
COMMUNICATION LINE

FIGURE 3

Prior to the offshore deployment trials at the Saltire location a complete inshore water trials deployment will be conducted. This is to gain confidence and experience in the system before the offshore trials are commenced.

3.0 COMMERCIAL JUSTIFICATION

The SPARCS system shows the greatest economic savings when used for single satellite subsea developments. A range of 10km is used as the maximum reliable distance for acoustic communications. The future development of other techniques not requiring umbilicals for communications will further improve the commercial viability of SPARCS.

The SPARCS system may also show economic advantages when used for multiwell and manifold type developments, but as these scenarios are dependant on factors outside the scope of this proposals, they are not considered here, and should be evaluated on a case by case basis.

1) The capital costs for an existing technology hydrocarbon field development is defined by the following item costs.

Note: Items common to both conventional and proposed technology installations are not included.

		Cost UK£000's
a)	Platform control system hardware for conventional installations (Hydraulic & Electrical power units)	300
b)	Subsea deployed control system hardware	250
c)	Umbilical manufacture cost (UK£180 per metre)	(180)x
	Where x is the umbilical distance is km.	
d)	Umbilical Deployment/Installation cost (UK£100 per metre)	(100)x
	Capital costs (UK£000's) = (280) x + 550	(i)

2) The capital cost for the proposed SPARCS technology not requiring umbilicals is;

		Cost UK£000's
e)	SPARCS Subsea Deployed Control System Hardware	1200

Note: This cost is not effected by variations in Platform/Wellhead distances up to 10Km.

3) The number of subsea development applicable to SPARCS is estimated at 5 off per annum averaged over a 10 year period.

4) The average distance from platform to wellhead, for the above field developments, where SPARCS is applicable, is estimated as 7.5Km.

2.0 PROJECT OBJECTIVES

2.1 Overall Aims - Cost Savings

The aim of this project is to reduce capital outlay, installation and maintenance costs of subsea field developments in relation to hydrocarbon production. This will be achieved by using an autonomous control system not requiring expensive umbilicals. The umbilicals are costly to specify and manufacture, require considerable expense and time to install and commission, and are relatively unreliable with the associated costs of maintenance and lost production.

The subsea deployed hardware for the SPARCS autonomous system is more complex and costly compared to the subsea hardware of a traditional umbilical system, but the overall costs of an autonomous system is comparable when the umbilical lengths total 2.3Km and shows increasingly better economics with longer wellhead to platform distances.

With the use of future technologies for non-umbilical communication techniques the maximum range of autonomous control system communications is expected to be in the range of 25 to 30 Km. Even at distances greater than 2.3Km the economic viability of field developments using autonomous systems is easily justified, as shown in section 3.0.

Another secondary aim of the project is to prove the technology of individual elements of the equipment as stand alone products. For example Subsea electrical power generation applications exist for remote telemetry and long distance communication repeater stations.

2.2 Applications Trials

The offshore application trials of the SPARCS system will demonstrate that an autonomous control system is able to perform reliable long term monitoring and control functions, in order to produce subsea hydrocarbons without the use of umbilicals.

The demonstration will prove and qualify all aspects of the system under realistic operating conditions with deployment on a typical subsea well in the North Sea. EE Caledonia have made available the Saltire WHPU water injection well No. 15/17-22Z with an acoustic link to the Saltire platform. A significant amount of subsea hardware (estimated at £50k) has already been deployed in anticipation of accepting the SPARCS subsea hardware.

The SPARCS demonstration system will be used to control the water injection well using acoustic telemetry over a representative distance of 2Km.

During the demonstration two forms of subsea electrical power generation will be evaluated, the first will be a rotating turbine Turbo Electric Generator (TEG) fitted to the Saltire water injection flowline, this will produce the actual power to operate the SPARCS subsea system. A second source of power will be produced from a Thermo Electrical Generator (THEG) attached to a nearby flowline containing high temperature production fluid. The THEG uses discrete solid state semi-conductor elements attached to the flowline and conducting thermal energy to the cold surrounding sea water to produce electrical power.

This second source of power would be connected to a dummy load and only monitored to evaluate long term performance. It would not be connected to the SPARCS system to produce power in this case, but used when controlling a production well.

Comparison with the production cost of conventional installations

1) The capital costs to develop 5 hydrocarbon subsea fields with convention technology and Platform/Wellhead distance of 7.5Km is:-

from (i) Capital (UK£000's) = 280 (7.5) + 550 per development

for 5off development Capital (UK£000's) = £2,650 x 5

= UK£13.25 million

2) The capital cost to develop the same 5 field using SPARCS is:-

Capital (UK£000's) = 1200 x 5

= UK£6.0 million

3) The annual cost saving using the SPARCS system is therefore UK£7.25 million.

4) The break even umbilical distance between SPARCS and conventional technology capital costs is 2.3Km.

Additional Recovery of Reserves

The aim of SPARCS is two fold, the first is to reduce capital and installation costs for subsea hydrocarbon developments where existing technology can be replaced by SPARCS technology, the economics of which are given in the above section.

The second aim is to allow recovery of subsea reserves which are not economically viable due to the high umbilical costs of existing technology. It is estimated SPARCS would allow hydrocarbon recovery of £87 million per annum over a period of 10 years based on the following assumptions.

1) Over a period of 10 years it is estimated 20 subsea field developments will become viable with SPARCS, which due to the high capital costs of existing umbilical systems would not be economic propositions.

2) The average of 2 fields a year will be relatively small in size, and produce an average of 10,000 barrels/day of oil for five years without gas lift assistance. This produces 7.3 million barrels of oil per annum.

3) The value of the above oil at $19/barrel and $1.6/£ is £86.7 million per annum.

The capital costs of the SPARCS subsea control system is estimated at £1.2 million per development, excluding platform facilities and hydrocarbon flowlines.

Maintenance costs for a SPARCS system

1) The annual operating and maintenance costs of the SPARCS system is in general the same as a conventional system.

2) The maintenance period for the SPARCS system is every 2 years with a total cost of £0.74 million for recovery, maintenance and re-deployment. The annual average cost is therefore £0.37 million per SPARCS system.

4.0 PROJECT PROGRAMME

4.1 Overall System

The SPARCS project has already completed the main research elements of the system. This has produced and demonstrated the technology in respect of electrical power generation, hydraulic power generation using subsea rated components and shallow water long distance acoustic telemetry.

The overall system concept has been thoroughly researched and shown to be a technically viable system.

4.2 Completed Work

The programme of completed work is shown in figure 4 and is summarised as follows:

Phases 1 to 3 defined the concept design and produced Functional Design Specifications for the overall system and main sub-systems. Phase 4A researched and developed prototype hardware for the Subsea HPU, Thermo-electrical Generator and power conditioning/motor control electronics.

4.3 Proposed Scope of Work

The proposed project consists of a further four phases, with a time schedule as shown in Figure 5:

Phase 4b - Hardware Design, Manufacture and Test
Phase 5 - Inshore Water Trials
Phase 6 - Inshore Water Trials Evaluation & Recommendations
Phase 7 - Saltire Offshore Deployment Application Trials

Phase 4b - Hardware

This phase completes the detail design, manufacture and land based testing of the outstanding hardware items.

Phase 5 - Shallow Water Field Trials

This work package allows for shallow water trials to be completed at a suitable inshore site. Portland Bay or Morecombe Bay are proposed as possible suitable sites in order to communicate over a long range (10Km).

Phase 6 - Evaluation

An evaluation of the work completed to date and detailed analysis of the shallow water field trials will be conducted in this phase. A series of recommendations in respect of the experience gained and possible impact on the operational requirements of the Saltire offshore application trials will be made.

Phase 7 - Application Offshore Deployment Trials

This phase consists of handover of the complete system to the offshore installation team, deployment and commissioning and system monitoring over a period of 18 months. This period is considered a minimum in order to prove the reliability of the acoustic communications over the seasonal weather condition variations.

Figure 4

SPARCS - PROGRAMME OF COMPLETED WORK
CONCEPT DESIGN & LAND BASED COMPONENT TESTS
PHASES (1,2,3 & 4A)

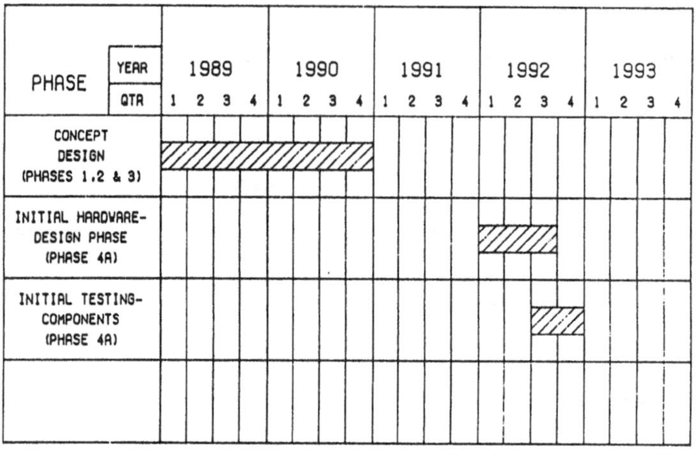

PHASE	YEAR	1989				1990				1991				1992				1993			
	QTR	1	2	3	4	1	2	3	4	1	2	3	4	1	2	3	4	1	2	3	4
CONCEPT DESIGN (PHASES 1,2 & 3)																					
INITIAL HARDWARE-DESIGN PHASE (PHASE 4A)																					
INITIAL TESTING-COMPONENTS (PHASE 4A)																					

Figure 5

SPARCS - PROGRAMME OF PROPOSED WORK
HARDWARE DESIGN, INTEGRATION & TEST (PHASE 1 & 2)
INSHORE TRIALS (PHASE 3)
OFFSHORE ASSEMBLY/COMMISSIONING & MONITORING (PHASE 3,4 & 5)

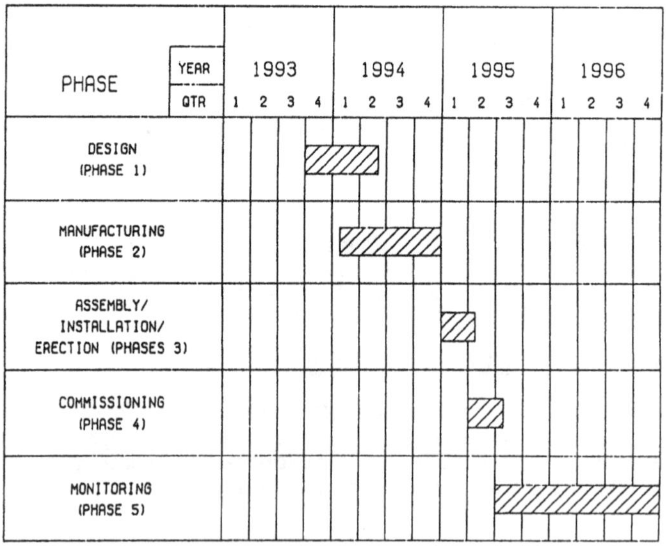

PHASE	YEAR	1993				1994				1995				1996			
	QTR	1	2	3	4	1	2	3	4	1	2	3	4	1	2	3	4
DESIGN (PHASE 1)																	
MANUFACTURING (PHASE 2)																	
ASSEMBLY/ INSTALLATION/ ERECTION (PHASES 3)																	
COMMISSIONING (PHASE 4)																	
MONITORING (PHASE 5)																	

4.4 Offshore Installation - Platform Equipment

The following equipment will be installed on the Saltire Platform.

1) Acoustic Transponder located on jacket leg A4. It is proposed the transponder is clamped and bolted, although consideration will be given to a wireline system to allow recovery to platform deck level.
2) Operator Control console installed in the Equipment Room.
3) Power and communication cable installed between transponder on jacket leg and Operator Control console (via Acoustic Modem).
4) Platform Power Supply for Operator Control Console and acoustic system.
5) Interface cable to Platform Emergency Shutdown system.

4.5 Offshore Installation - Subsea Equipment

The following equipment will be installed on the Saltire WHPU subsea structure.

1) Acoustic Transponder
2) SPARCS Subsea Control Module
3) Turbo Electric Generator (TEG) installed in water injection flowline and connected to SPARCS Subsea Control Module.
4) Pressure sensors and associated subsea cables and connectors.
5) System switch intervention unit to enable well control to be operated via direct hydraulic platform link or SPARCS.
6) Thermo-Electric Generator (THEG) installed on a nearby production well flowline with a performance data monitoring line connected back to the Saltire platform to evaluate long term performance.

5.0 SYSTEM DESCRIPTION

The Subsea Powered Autonomous Remote Control System (SPARCS) will have the ability to control a single remote water injection or production well. The system will comprise of two groups of components, the Surface Controller and the Subsea Control System. The Surface Controller forms the control station which provides the operator with the means of controlling and interrogating the subsea system. The Subsea Control System forms the outstation which provides all the control and data monitoring functions of the well.

The use of new technology such as electrical actuators or sea water actuators will be evaluated when qualified actuators are available.

As the SPARCS system is autonomous and generates power subsea, there will be a reduction in the amount of surface equipment required at the platform facility for electrical and hydraulic power units as used in traditional control systems. Removing these equipments from the potentially hazardous environments found on offshore platforms, will result in increased safety levels for personnel and operations.

5.1 Surface Controller

The surface controller comprises:-

a) Operator Control Console

b) Surface Acoustic Transponder

The operator control console has the operator keyboard, the display monitor, a printer and a functional control panel station. The computers are rack mounted and comprise, CPU, memory, I/O cards, disc storage and the computer watchdog.

The Surface Acoustic Transponder will house the Communication Modem, the Acoustic Transmitter-Receiver and interface to a directional hydrophone. It will be connected to the leg of a platform, or a separate cage frame on the seabed, and directed towards the Subsea Control System.

5.2 Subsea Control System

The Subsea Control System comprises:-

a) Turbo Electric Generator (TEG)
b) Submersible Battery System
c) Power Conditioner System
d) Motor Controller System
e) Subsea Electronics Module
f) Hydraulic Power Unit
g) Subsea Control Module
h) Subsea Acoustic Transponder
i) Subsea Control Valves
j) Thermo Electric Generator (THEG)

The Turbo Electric Generator (TEG) will convert the kinetic energy of a injection water flowline into electrical energy to power the Subsea Control System, when controlling a water injection well.

The Submersible Battery System will store the electrical energy from the TEG in periods of low power demand and will provide power to meet the peak demands for electrical power.

The Power Conditioner System will interface the TEG and the electronics/batteries. It provides power to the electronics plus it maintains the batteries charge.

The Motor Controller System provides the interface between the system power and means of motor control, and the motors.

The Subsea Electronics Module contains the electronics for controlling and monitoring valves and data.

The Subsea Hydraulic Power Unit is required to provide the hydraulic fluid power for the valve control system and comprises reservoir, pumps, motors, accumulators and filters. Dual high pressure supply lines are available.

The Subsea Acoustic Transponder will house the communication modem, the Acoustic Transmitter/Receiver and interface a directional hydrophone. It will be connected to (or near) the wellhead Subsea Control System and directed toward the Surface Acoustic Transponder, ensuring a clean line of sound.

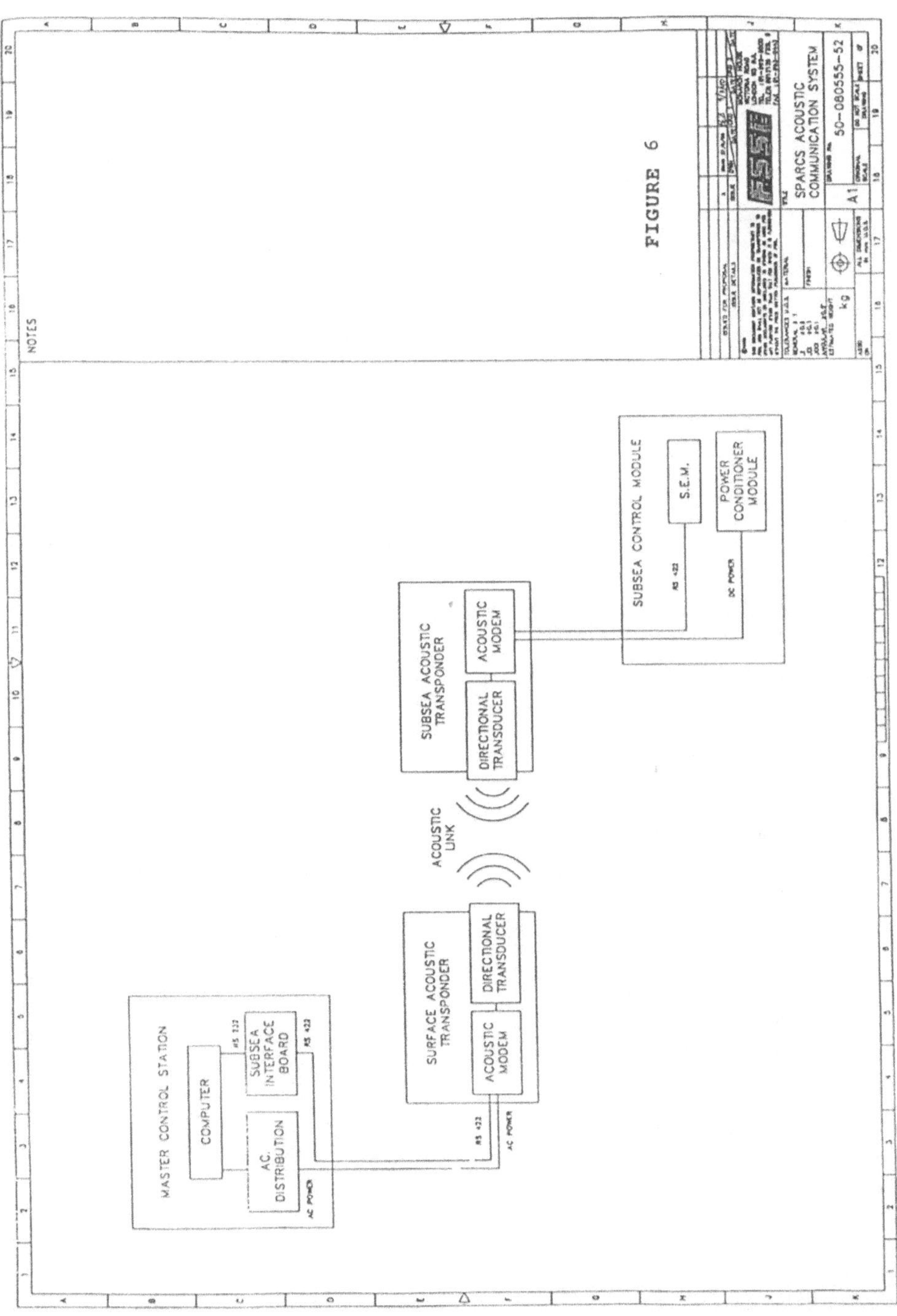
MASTER CONTROL STATION
COMPUTER
SUBSEA INTERFACE BOARD
A.C. DISTRIBUTION
RS 232
RS 422
AC POWER
SURFACE ACOUSTIC TRANSPONDER
ACOUSTIC MODEM
DIRECTIONAL TRANSDUCER
ACOUSTIC LINK
SUBSEA ACOUSTIC TRANSPONDER
DIRECTIONAL TRANSDUCER
ACOUSTIC MODEM
SUBSEA CONTROL MODULE
S.E.M.
POWER CONDITIONER MODULE
RS 422
DC POWER
NOTES
FIGURE 6
SPARCS ACOUSTIC COMMUNICATION SYSTEM
50-080555-52
A1
kg

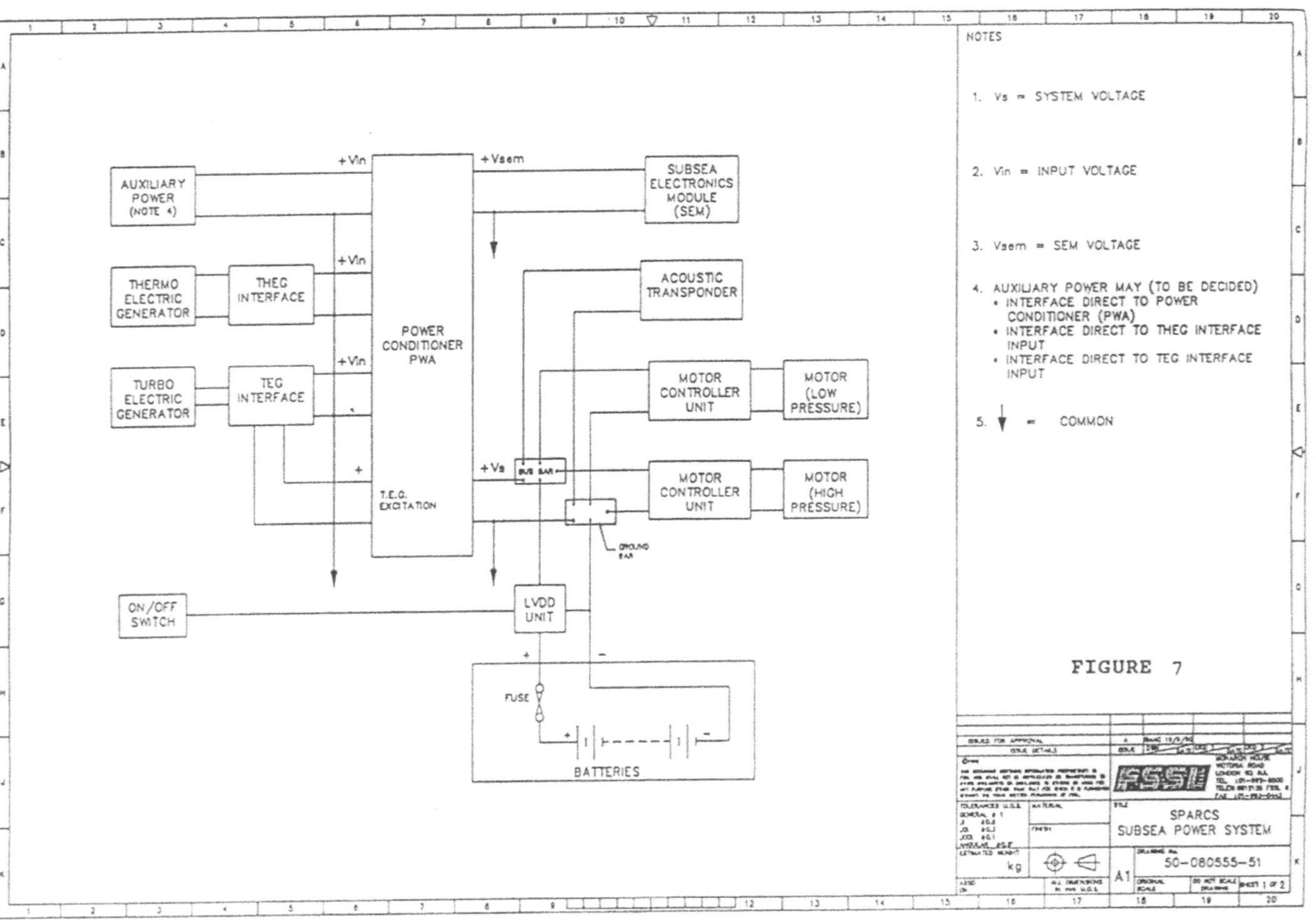
AUXILIARY POWER (NOTE 4)
THERMO ELECTRIC GENERATOR
THEG INTERFACE
TURBO ELECTRIC GENERATOR
TEG INTERFACE
+Vin
+Vsem
+Vs
POWER CONDITIONER PWA
T.E.G. EXCITATION
SUBSEA ELECTRONICS MODULE (SEM)
ACOUSTIC TRANSPONDER
MOTOR CONTROLLER UNIT
MOTOR (LOW PRESSURE)
MOTOR (HIGH PRESSURE)
BUS BAR
GROUND BAR
ON/OFF SWITCH
LVDD UNIT
FUSE
BATTERIES
NOTES
1. Vs = SYSTEM VOLTAGE
2. Vin = INPUT VOLTAGE
3. Vsem = SEM VOLTAGE
4. AUXILIARY POWER MAY (TO BE DECIDED)
• INTERFACE DIRECT TO POWER CONDITIONER (PWA)
• INTERFACE DIRECT TO THEG INTERFACE INPUT
• INTERFACE DIRECT TO TEG INTERFACE INPUT
5. = COMMON
FIGURE 7
SPARCS
SUBSEA POWER SYSTEM
50-080555-51

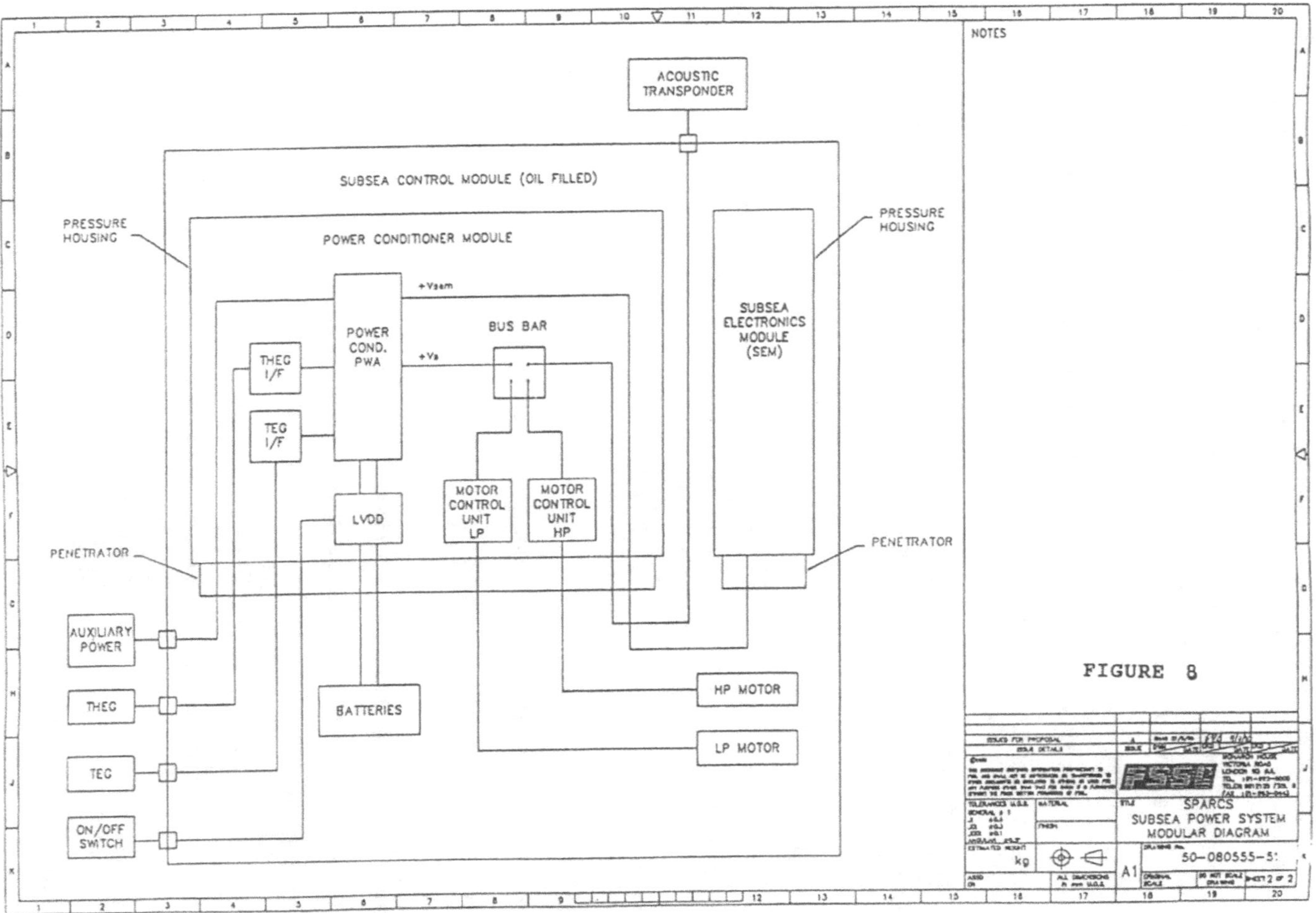
NOTES
ACOUSTIC TRANSPONDER
SUBSEA CONTROL MODULE (OIL FILLED)
PRESSURE HOUSING
POWER CONDITIONER MODULE
+Vsem
POWER COND. PWA
+Vs
BUS BAR
THEG I/F
TEG I/F
LVDD
MOTOR CONTROL UNIT LP
MOTOR CONTROL UNIT HP
SUBSEA ELECTRONICS MODULE (SEM)
PRESSURE HOUSING
PENETRATOR
PENETRATOR
AUXILIARY POWER
THEG
TEG
ON/OFF SWITCH
BATTERIES
HP MOTOR
LP MOTOR
FIGURE 8
SPARCS
SUBSEA POWER SYSTEM
MODULAR DIAGRAM
50-080555-5:
A1
kg

The Subsea Control Module (SCM) supports the above subsystems in making up the entire Subsea Control System. It is the SCM that is mounted on the support frame near the wellhead. The Solenoid Control Valves are the electrically operated control valves used for actuating the Tree Valves, and are contained within the SCM.

The Thermo Electric Generator (THEG) will be mounted on a nearby production facility. It will be interfaced to SPARCS solely for experimental evaluation and will provide power if SPARCS is required to control a production well.

5.3 General Well Data

Single water injection well located between 2 - 10 km from the surface facility.

Completion Type: 5" and 2" dual bore
Design Pressure: 517 bar

Surface Environmental Data

Operating Temperature:	5ºC to 30ºC
Design Temperature:	0ºC to 45ºC
Room Pressure:	25 - 65 Pa above atmospheric
Humidity:	20 - 80% non condensing

Subsea Environmental Data

Depth:	50 to 250 meters
Current:	Seabed 0 to 0.5m/s
Temperature:	Surface 3ºC to 16ºC
	Seabed 4ºC to 10ºC
	Air -10ºC to +25ºC
Density:	1.0233 to 1.0292 g/cm^3
Salinity:	Typically 3.5%

Well Valve Data

The water injection well has the following complement of valves:-

Valve	Size
Tubing Master	5 1/8 inch
Tubing Wing	5 1/8 inch
Tubing Choke	4 inch
Subsea Safety Valve (SSSV)	5 1/2 inch

The valves have the following characteristics:-

5 1/8 inch valves:-

Swept Volume:	3.2 litres
Typical Open Pressure:	54 bar
Typical Cracking Pressure:	47 bar
Maximum Working Pressure:	207 bar

SSSV Tubing valve:-

Swept Volume:	0.09 litres
Typical Pressure:	517 bar *
Typical Cracking Pressure:	To be advised

* At 345 bar well pressure

Choke valves:-

Body Pressure		345 bar
Actuator Range	0 - 100%	0 - 90 steps
Displacement		1.59 litres
Actuator Pressure		57 bar

Hydraulic Control Functions

The Tree Valves to be controlled are Master, Wing, SSSV, Choke Valve Open, Choke Valve Close with operating pressures up to 517 bar. A sixth spare control function is to be implemented. (Choke valves are only used on production wells, not implemented for Saltire trials).

Well Sensor Data

The water injection well has the following data sensors:-

Tubing Pressure	0 - 400 bar	4-20mA
Annulus Pressure	0 - 400 bar	4-20mA
Choke Differential Pressure	0 - 20 bar	4-20mA
Choke Position	0 - 100%	4-20mA

The rate at which these data points are scanned and reported to the surface is shown in Table 1. Flow for production wells can be calculated utilising the differential pressure across the choke and the valve's Cv position curve. For water injection wells, flow can be determined from TEG output voltage.

Facilities will be made, if practical, to monitor downhole pressure and temperature from a nearby producing well. Typical downhole pressures and temperatures are up to 690 bar and 150°C.

Valve Operating Frequency

There are six modes (out of eight) in which a water injection well can be operated, and for each mode there is a set status for the well valves. The well valve status for each well mode is shown in Table 2. The number of times that well mode is selected is dependant upon the operating conditions (Maintenance, normal injection or Dormant).

5.4 Hydraulic Fluid

The SPARCS control system is to use a water based fluid which is compatible with the Saltire system fluid (HW540) but has a higher viscosity similar to that of a mineral oil based fluid.

Oceanic HV355 will be used as the control fluid. It has the following values:-

Kinematic Viscosity	60 cs at 0ºC to 29 cs at 20ºC
Density	1065 kg/m^3

The suppliers (Marston Bentley) have confirmed it is chemically similar to HW540, and is fully compatible.

5.5 Design Philosophy

General

The overall design philosophy is to produce an autonomous subsea control system which conforms to the functional requirements, engineering standards and safety codes of practice of existing control systems. Being an autonomous system particular regard is to be made to its safe operation, and any single point failure should not leave the well in an unsafe condition.

Due to the difficulty and cost of access to subsea systems the design of the control system shall result in high reliability and availability. The system is to be inherently reliable, using components of high quality, and where possible of proven or field tested history.

The target for availability of the SPARCS system is 98% with a down time of 10 days in a year. These figures may not be easily verified, since reduced availability could be due to environmental conditions beyond operational control (eg, acoustic disturbances).

The materials of the SPARCS system shall be selected to resist deterioration during long term immersion in seawater and in contact with the operating well and hydraulic fluids. The design shall minimise the possibility of electrolytic action due to dissimilar metals. Cathodic protection systems are to be developed in conjunction with the protection system used on associated wellhead subsea equipment.

The design life of the subsea portion of the SPARCS system is to be 20 years with periodic maintenance. For this application on a water injection well, the system shall be designed to operate for a minimum period of at least 2 years without diver intervention for maintenance. It is proposed that the system will be tested and operated in two phases;

(1) Shallow water trials - 3 to 4 months continuous operation
(2) Application trials - up to 18 months operation subsea

5.6 Installation and Maintenance

Installation Methods

SPARCS shall be designed to reduce subsea installation time and costs. The subsea control system is to be mounted on a support frame. The support frame shall have provisions for being guided to and located on a standard permanent guide base (PGB). This control system guide base is to be attached to a well conductor which is drilled and set into the seabed within close proximity to the well which it is to control, or a guide base which is part of a subsea production template. The components and sub-systems of the control system are to be integrated as much as is practicable so as to minimise the number of wet mateable electrical connectors.

Proposed Well Sites

For the initial application of SPARCS, the initial installation will be in parallel with the existing control system of the Saltire WHPU single, Water Injection Well No. 15/17-22Z. This site has the capability to replenish hydraulic fluid reserves without intervention, therefore making this site ideal for the demonstration trials.

The design shall provide the capability of changing back to the existing control system quickly, in the event of acoustic system failure. Connection from one control system to the other may be either:-

a) Diver mateable connections for both hydraulics and sensors.
b) ROV operated valves for hydraulics and ROV operated switches for sensors
c) Remotely operable hydraulics.

Maintenance

The period between planned maintenance is to be the same as that for well maintenance. In general for an injection well this will be 2 years. For this application on the proposed water injection well, the system shall be designed to operate for a minimum period of at least 24 months.

Maintenance of the control system is to be achieved by diver intervention. Maintenance will be affected by retrieval of the support frame which locates the subsea control system to the surface where maintenance is carried out. The Permanent Guide Base (PGB) would not be retrieved. Hydraulic fluid replenishment will be completed without the need for retrieval of the subsea equipment. The components of the system that require maintenance are:-

Hydraulic	
Replenish hydraulic fluid	2 yrs
Replace filters	5 yrs
Maintain/replace motor pump sets	5 yrs
Re-charge accumulators	5 yrs
Electrical	
Replace batteries	2 yrs
Acoustic	
Maintain transponder	5 yrs
Turbo Electric Generator	
Turbine, Bearings, Seals	5 yrs

5.7 Power Generation

For the application on the Saltire Injection Well, a Turbo Electric Generator (TEG) only, will be used for SPARCS power. The Thermo Electric Generator (THEG), which is not applicable to injection wells, will be installed nearby and monitored.

The Turbo Electric Generator (TEG) is to be installed in the flow line spool of a water injection well. It will convert the kinetic energy of the injection water into electrical energy providing continuous power to the entire Subsea Control System. In general the TEG will provide 100 watts minimum of output power. It shall operate to 250 metres depth in a seawater environment. The output of the TEG will be 3 phase low voltage AC (11 VAC to 20 VAC). The output voltage of the TEG may also be exploited as means of measuring water flow in the pipeline.

The Thermo Electric Generator will rely on the temperature drop between the hot produced fluids of a production well and the cold ambient seawater to generate power. The experimental THEG will provide upto 20 watts of power. It shall operate to 250 metres depth in a seawater environment. The output of the THEG will be a low voltage DC (3 VDC to 8 VDC). Nominal power of the THEG when fitted to a production flowline will be 100 watts.

5.8 Acoustic Transponder (Surface and Subsea)

The surface and subsea acoustic transponders are required to carry out all communication functions of SPARCS. The units will provide secure acoustic communication using a semi-duplex serial link with error detection and correction facilities. In normal use, the subsea unit will act as a slave to the surface equipment. However, under certain circumstances the subsea unit may initiate a transmission to signal fault conditions detected within the subsea system.

System Range:	10Km
Data Resolution:	12 Bit
Host I/O:	RS422
Low Power Consumption:	200 watts maximum
Target Error Rate:	Better than 1 in 10^4
Maximum Operating Depth:	250 metres
Data Rate:	Better than 40 BAUD

The 2 off Acoustic Transponders (1 wellhead/1 platform) will consist of:-

- Directional Transducer (Hydrophone)
- Transmit Unit
- Receive Unit
- Transit/Receive Switch
- Acoustic Modem with RS422 I/O to Operate Console & SEM
- Local Power Supply Conditioning

Figure 5 shows a block diagram representation of the SPARCS acoustic communication system.

5.9 Hydraulic Power Unit

The Hydraulic Power Unit will provide hydraulic power at two pressure levels. The hydraulic system is to be a closed loop system where all leakage and venting flow is returned to the reservoir. The reservoir is to be fully pressure compensated to sea depth pressure.

The use of return line surge accumulators is to be used to limit the peak flow into the reservoir/receiver system. The reservoir receivers shall have provision for the separation of gas and solids from the return flow and means of venting gas pressure build up. It is essential that an impermeable interface is introduced between the hydraulic oil and the pressure compensating fluid. The reservoir is to be sized to supply all the tree actuators and supply accumulators at once with an allowance for irrecoverable hydraulic fluid leakage over the maintenance period. The suction line shall incorporate a suction strainer to prevent large particular contamination from reaching the pump.

The pumps are be to submersible d.c. motor driven close coupled pump sets. The motors will be powered from the Motor Controller System. Both motors and pumps are to be immersed in an oil filled and pressure compensated environment.

Receiver capacity	- 100 litres
Reservoir capacity	- 300 litres
Reservoir vent pressure	- 0.5 bar above sea pressure
Pressure nominal	- 517 bar - HP (High Pressure)
Pressure nominal	- 207 bar - LP (Low Pressure)
Suction line strainer	- 100 u
Flowline filtration	- 3 u

Motor pump sets:-	
Voltage	- nominal 60 Vdc
Output Power	
(70% efficiency) LP	- 350 watts minimum
(70% efficiency) HP	- 300 watts minimum
Flow LP	- 0.7 litres/minute minimum
Flow HP	- 0.2 litres/minute minimum

The SPARCS system uses a closed loop hydraulic system which does not dispense control fluid to the sea during well valve operations. Although the majority of fluids are water based the use of closed loop systems is inherently of less impact to the environment. A considerable amount of this fluid is used in offshore hydrocarbon production. It is estimated the SPARCS system will reduce pollutant emission by approximately 3000 litres per annum.

5.10 Submersible Battery System

The submersible battery system is required to provide back up power or energy for periods of peak power demands. It will be installed and operated (including charging) subsea in an oil filled environment at depths up to 250 metres. In general a pressure equalised battery system using flooded liquid electrolyte lead acid cells will be used. The nominal battery system capacity will be 18,000 watt. hours at 20ºC. Nominal system voltage is to be 60 VDC with a maintained float charge. The submersible battery system will be integrated with the other components of the SPARCS System.

5.11 Power Conditioner System

The Power Conditioner System provides the interface between the subsea generators (TEG or THEG) and the Electronics/Battery System. It will power the Subsea Control System as a priority over all other system power requirements. It will also charge the battery system at float and higher charge rates.

It is proposed that a Power Conditioner circuit board along with the THEG Interface or TEG Interface and Low Voltage Detect and Disconnect (LVDD) unit be installed within a 1 atmosphere, nitrogen filled pressure vessel - called the Power Conditioner Module.

Figures 7 and 8 show a block diagram representation of the power conditioner system.

5.12 Motor Controller System

The Motor Controller System provides the interface between the batteries, and power source of motor control and the motors.

The batteries will provide the source of energy for the motors via motor power units. Motor control electronics will interface the system microprocessor and motor control pressure switches to the motors, enabling on/off control. It is proposed that the motor power units and motor control electronics will be installed within the 1 atmosphere, nitrogen filled pressure vessel, of the Power Conditioner Module.

5.13 Subsea Electronics Module (SEM)

Functional Requirements

The SEM is responsible for:-

- Acoustic communication control (subsea module).
- Mode Operation.
- Operating the hydraulic control valves (to open or close well valve actuators).
- Exciting and monitoring the system sensors.
- Monitoring various system voltages.
- Data logging of sensor and voltage information.
- Executive local control actions.
- Calibration check facility.
- Dealing with loss of communication.
- Motor control.
- Down loadable application software.
- Low power consumption (power down control).
- Quiescent receive mode

Acoustic Communication Control (Subsea Module)

The SEM will be responsible for controlling all subsea acoustic transponder transmissions and also operate in a quiescent receive mode awaiting transmissions from the surface acoustic transponder. It will also have the capability to wake up the Subsea Electronics Module. The electrical communication protocol between the SEM and subsea acoustic transponder is to be over a RS 422 serial link.

Mode Operation

The water injection well will be operated in six (out of eight) operational well modes.

These are:-
1) Field Well Shut in class 1
2) Process Well Shut In class 2
3) Well Start Up
4) Normal Injection
5) Reserved for Production Well
6) Reserved for Production Well
7) Platform Manual Override
8) Well Dormant

The principle behind mode operations is that since acoustic communication is slow (propagation and baud rate), a single mode command from the surface defines a complete tree configuration, replacing several separate valve commands. Additionally, mode operation offers a saving in subsea power, since the subsea control module needs to acknowledge only the one mode message. Single valve operation is required (Mode 7).

Each operating mode has particular requirements for valve functioning (table 1).

Table 1

VALVE FUNCTIONING AND SEQUENCIES (WATER INJECTION WELL)

Well Valve	Mode No.	1		2		3		4		7		8	
	permissiv-ness	2,3,4,5 6,7,8		3,4,5,6 7,8		1,2		3,		1,2,3,4,5 6,8		1,2	
	Well Mode	Field Well Shut in		Process Well Shut in		Well start up		Normal Injection		Platform Manual Override		Well Dormant	
		Seq No.	Final Pos tn	Seq No.	Final Pos tn	Seq No.	Final Pos tn	Seq No.	Final Pos tn	Seq No.	Final Pos tn	Seq No.	Final Pos tn
Tubing SCSSV		3	C	-	O	1	O	-	O			-	O/C
Tubing Master		2	C	-	O	2	O	-	O			-	O/C
Tubing Wing		1	C	1	C	3	O	-	O			-	C
Tubing Choke		-	-	-	-	-	O	A	O			-	-

Key:
- C = Valve Close
- O = Valve Open
- O/C = Valve left as either Open or Closed
- \- = Valve operation inhibited
- A = Operate as required

Permissiveness - This mode can only be entered from permitted modes

5.14 Technical/Financial Risks

Technical risks:

Technical risks stem from deploying equipment subsea without a previous history in this environment. The offshore operation phase of this project will qualify the design fully to allow future commercialisation.

Effects which contain risk for the Thermo-electric generator are the long term exposure to sea water and marine growth which could cause electrical short circuits through water ingress. Failures may also occur as a result of corrosion or an unacceptable reduction in power output, due to marine growth preventing sufficient thermal gradients.

There is considerable technical risk in the integration aspects of this system where the overall power balance could result in consumption exceeding available energy, and the gradual depletion of stored battery power. This situations is dependant on hydraulic fluid loss leading to an excessive duty cycle of pump/motors to maintain system hydraulic pressure.

Financial risks:

As the supply of umbilicals to the oil industry is now an established market, it is not estimated that any significant cost reductions in umbilicals will be possible. The financial viability of the SPARCS system is therefore not sensitive to this situation.

The estimates for the number of offshore hydrocarbon field developments, in the next 10 years, is considered to be on the conservative side. Any increase in these estimates will improve the probability for commercial success of the SPARCS system.

Financial risks also involve the SPARCS system not operating with the estimated reliability, and an associated loss of hydrocarbon oil revenues during maintenance operations. These risks will be minimised by demonstrating and evaluating the design during the application trials in realistic operating conditions. This will present an opportunity to incorporate any relevant modifications into the system design prior to the product launch.

6.0 ACKNOWLEDGEMENTS

The author wishes to thank Enterprise Oil Plc and Elf Enterprise Caledonia Ltd for their support and help in progressing the SPARCS project.

Session 3
Project Management and Risk Minimisation

IMPACTS OF WEATHER AND CLIMATE ON THE OFFSHORE INDUSTRY

J.C. THOMSON
Oceanroutes (UK) Ltd
Swire House, Souter Head Road, Aberdeen AB1 4LF
Scotland

ABSTRACT The use and application of meteorological data in support of the offshore oil and gas can assist in all aspects of operations. From the use of climatological data for pre-operational planning or for design purposes through to real-time weather forecasts for ongoing and weather-sensitive operations, the application of reliable and accurate meteorological data can have a significant impact on safety and cost-effectiveness.

INTRODUCTION

Weather and climate have a major impact on many aspects of life, and perhaps more so in the harsh environment of the North Sea than in many other spheres.

Whilst "weather" is usually taken to refer to those conditions which may be expected over the next few days, "climate" is generally assumed to mean the average conditions which occur over a long number of years - and hence the conditions which are likely to be met in any succeeding year. This latter definition will only hold true if we assume that the climate is not undergoing a significant change, but even so, climate change is generally a gradual process and hence statistical data derived from past years of historical data are still likely to give an adequate indication of conditions which can be expected.

CLIMATOLOGICAL DATA

In determining, for example, the times of year when a specific offshore operation can be undertaken, or the optimum time for performing this operation, climatological data is frequently used. Various databases of marine operations are available based on observations taken by merchant ships, and often presented in the form of summarised monthly statistics for a particular area. It is important to note that the data are not specific to a location but are gathered over sometimes quite a wide area and hence some care should be taken in their use.

A typical monthly summary might contain tables of average wave height against wave period, wind speed against direction, etc., and are certainly of use in the initial planning of an operation. Again, because we are dealing with ship data some caution is advised, since, when possible, ships will avoid areas of severe weather and

Volume 30: Subsea International '93, 151–157.

consequently there is at least an implied bias towards fair weather conditions in the summarised data.

Many environmentally related problems require a time-continuous database for achieving an optimum solution. From a time-continuous database it is possible to determine such information as:-

a. The total amount of downtime to be expected for an activity or series of activity.

b. The probability of an operation being completed within a specified time period.

c. The optimum sequence of activities to take advantage of likely weather conditions.

d. The persistence of environmental conditions such as wave height or wind speed to aid in fatigue/stress analysis.

Since long period wave records are rarely available, it is necessary to use a different approach to obtain a time-continuous database of wave parameters. A commonly used time-continuous database is the Global Spectral Ocean Wave Model (GSOWM) produced by the United States Navy. This model starts with wind fields, which can be readily derived from historical weather charts, and utilises these as inputs to a spectral wave prediction model with a time resolution of 12 hours.

Climatological data is an important element not only in planning but also in the design of offshore structures. By applying standard statistical techniques to climatological data it is possible to determine the 50-year wave which is then incorporated in the design phase. Clearly the calculation of the return wave is important, since under-design could compromise safety while over-design would lead to unnecessary costs.

But is climate really changing and if so what are the implications? Certainly climate has changed over the centuries, but perhaps these can be considered as fluctuations about some "mean" climate. The earlier held belief by meteorologists that 30 years of data were sufficient to define climate for all time is no longer valid and the current view is that there can be significant climate changes over timescales of decades. But what implication is there in this as far as the North Sea offshore industry is concerned.

Recent work by Bacon and Carter (1991) on wave climate changes in the North Sea and the North Atlantic point to an increase in mean wave height over the whole of the North Atlantic in recent years, possibly of as much as 2 per cent per year. Bason and Carter point out, however, that "there are insufficient data to say with any confidence whether maxima or extremes have also risen".

In the North Sea the data would imply that the mean wave height increased from about 1960 to a peak around 1979-1980, with the mean wave height for 1984 about 13 percent below the peak value. Bacon and Carter, however, go on to point out that "recent winters (particularly 1988-1989) have produced severe conditions in the southern Norwegian Sea and northern North Sea", and that "without more data, it is not known whether the mean or the extreme climate is changing".

While it might be expected that the extreme values would increase with the means, Hogben (1989) has suggested that this is not necessarily the case and that the increase in the mean values could be attributed to an increase in the height of the swell and not in the locally generated wind wave and that in severe storms the swell component is negligible.

The question of how climate change might affect North Sea operations is not very clear, but if Hogben's suggestion is accepted, there would be little or no implication for design purposes. Perhaps, rather than take the traditional view that the longer the historical record the better will be the average statistical conditions it would be better to use the most recent database as the basis for the calculation of both average and extreme statistics.

FORECASTING FOR THE OFFSHORE INDUSTRY

If climatological data can perhaps be considered as a strategic approach to offshore operations then forecasting services can be considered as the tactical approach.

But how good is the forecast and what sort of reliability can the offshore industry place upon a forecast service? While there is undoubted agreement that there has been a substantial improvement in weather forecasting performance over the past decade or more, it must still be remembered that despite the advent of satellites and high speed powerful computers, weather forecasting is still a predictive science and as such will not always achieve complete 100 per cent accuracy.

What is right in terms of a forecast is perhaps not always easy to assess objectively. If a significant wave height of 3.0 metres is predicted but 3.5 metres occurs is the forecast wrong? While to the forecaster, a difference of 0.5 metre might well be considered within acceptable margins of variability, to the operator whose wave limit for a specific operation is 3.0 metres such a forecast may be perceived to be wrong.

Whilst statistical assessment of forecast accuracy may well show a "hit" rate of 90 per cent of more in the 12-hour and 24-hour period, as far as the offshore operator is concerned the forecast will fulfil its purpose if it allows him to perform - or correctly leads him to discontinue - a specific operation.

In the statistical approach to forecast assessment only the Canadian Oil and Gas Administration (COGLA) appears to have a rigourous approach to forecast assessment and requires all providers of forecast services to assess their forecast performance in accordance with the guide-lines laid down.

The COGLA guide-lines include the commonly-used approach of "hits" and "misses" but also incorporate a time element rather than rigourously insisting on "spot" values. This is a sensible approach since on occasions the error in a forecast may simply be an error of timing rather than magnitude of the waves or strength of the winds. In addition the bias and the mean absolute errors in the forecasts of wave height and wind speed are calculated. COGLA also requires that parameters relating to the forecasting and occurrence of severe weather are calculated, since it is considered important to assess those occasions when:-

a. Severe weather was forecast and occurred.

b. Severe weather was forecast and did not occur.

c. Severe weather was not forecast and did occur.

Clearly the aim is to achieve condition (a), since (b) will lead to false alarms and unnecessary costs and (c) to potentially dangerous and life-threatening situations.

It is all very well to talk about forecast accuracy and statistical evaluation of forecast, but such evaluations are only possible when adequate, reliable data is available from the location of interest. Not only is data absolutely essential for forecast assessment it is also a vital ingredient in the forecasting process. Whilst most platforms and rigs in the North Sea do perform meteorological observations, the quality of observations is variable and in many cases the frequency inadequate for a true statistical assessment of the forecasts. There are a number of platforms which have installed sophisticated instrumentation systems, taking measurements automatically every 10 minutes and making the data available in real-time to forecasting agencies; such data are invaluable in the forecasting process.

In weather forecasting, it is necessary in the predictive process to start with the latest possible data, hence the need for a good flow of data from the North Sea and much wider afield. An analysis - both of surface and upper air data - is the essential starting point of weather forecasting. I then becomes the forecaster's task - with the aid of computer-generated products - to determine how weather patterns are going to change over, say, the next 72 hours and impart this information to the client in such a manner that the information is both useful and incapable of being misinterpreted.

Forecasts have traditionally been provided to the offshore industry in the form of a lengthy telex message. It is fairly general that this telex contained both a text description of the expected conditions and a tabular section presenting point/time values of wind and wave parameters. It is always pointed out that the tabular section should be read in conjunction with the text section and not treated in isolation. Failure to do so has the potential to lead to misinterpretation of the conditions to be expected.

A typical tabular section of a forecast might contain:-

Date Time	12Sep 1200Z	12Sep 2359Z	13Sep 1200Z	13Sep 2359Z	14Sep 1200Z
Sig Wave (m)	3.5	4.5	5.0	3.5	2.0
Wind 10m (kts)	28	38	45	30	20

Such a table does not, for example, give any indication of when the highest wave height or greatest wind speed will occur of their values. Will the waves peak on the morning of 13 September or during the afternoon? From the above table there is no means of telling and hence the user would be unable to make the best possible weather-related decision, although a better decision is likely if the tabular section is used in conjunction with the text.

More recently there has been a move to presenting forecasts in graphical format, by utilising computer-to-computer transmission of forecast data. In some cases the graphical forecasts have been made available on a client's corporate computer network hence allowing access throughout the company on an international basis. In the latter approach, rather than transfer large graphics files via model to the host computer, data files can be transferred which are then converted into graphics images at the host. by adopting this method, transmission times and costs can be reduced substantially. As a back-up to data transfer by model and to pass graphical forecasts to offshore locations without the facility to receive or display graphical data on computer, facsimile can be used.

In particular, presenting wind and wave forecasts as 72-hour time series graphs makes misinterpretation of the forecast much less likely. Typical examples are shown in Figure 1 and Figure 2.

From the graphical presentation it is extremely simple to identify weather windows during which wind and wave conditions are unlikely to impede or delay the progress of the operation. For example, if an operation requires a period of 12 hours during which the maximum wave height must not exceed 3 metres, the likelihood of obtaining these conditions and how long they will last can be determined at a glance. It is equally simple to determine when conditions are forecast to be such that precautionary actions must be taken or a specific operation discontinued.

It is often the case that for particularly weather-sensitive operations such as well testing, subsea construction, etc., an offshore client will make use of an onsite meteorologist.

If climatological data is considered the strategic approach, and forecasting the tactical approach, then an onsite meteorologist represents the operational element in the application of meteorological data. The importance of the onsite meteorologist is in his ready availability to offshore management and his on-the-spot assessment of conditions, their duration and their likely changes. The main priority is "nowcasting", or the provision of short range weather forecasts for such weather-sensitive operations as off-loading a structure from a barge. The aim of the onsite meteorologist is to enable offshore management to make the maximum use of the available weather windows.

ECONOMIC BENEFITS OF WEATHER AND CLIMATE

It has been estimated there is a substantial cost-benefit in National Meteorological Services. Such authorities as Mason (1966), Tolstikov (1968) and Maumder (1972) have estimated a cost/benefit ratio of 1:20, 1:5 and 1:17 for the United Kingdom, the Soviet Union and New Zealand. More recently Maunder has estimated that in 1985 the benefits to agriculture, building and construction, and manufacturing in New Zealand total at least $80 million, resulting in a cost/benefit ratio in these sectors alone of 1:5. Whatever the accuracy of these figures, it does seem likely that a ratio exceeding 1:10 can confidently be assumed. As Mason (1966) says:

"The economic value of a special service for a particular customer can usually best be judged by the customer himself, but he may be reluctant to divulge this for the fear that the charges will be increased".

In some areas of commercial meteorology it is not too difficult to arrive at definitive

FORECAST WIND CONDITIONS
NEDDRILL 6 UK 21/4B

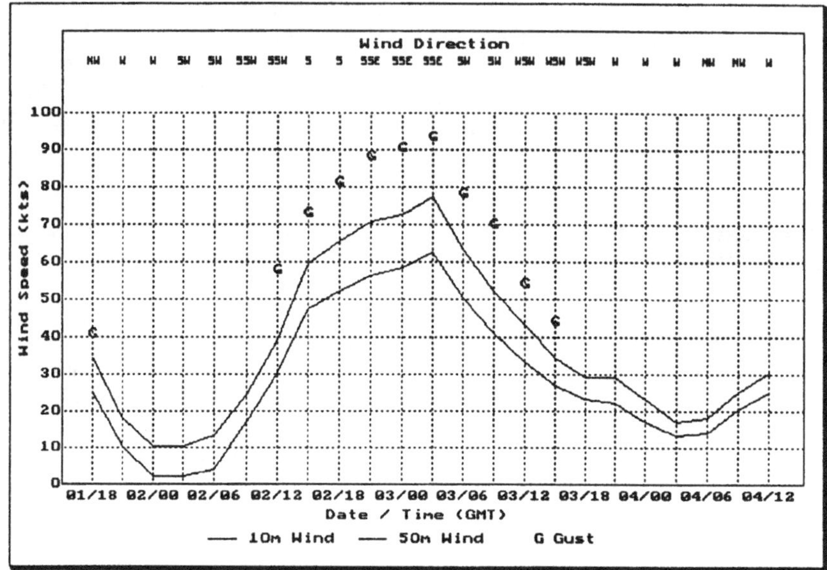

Fig. 1

FORECAST WAVE CONDITIONS
NEDDRILL 6 UK 21/4B

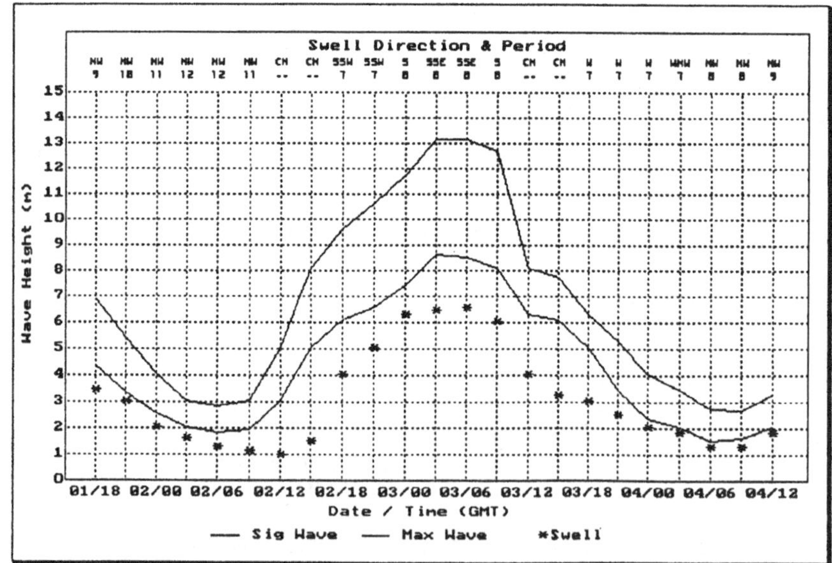

Fig. 2

answers in respect to the value of weather services. In shiprouting, for example, Motte and Calvert (1991) have estimated that on trans-Atlantic crossings modern vessels capable of 20 knots and using a shiprouting service, saved on average some 3-4 hours compared with unrouted ships. For the Pacific Ocean, the average time saved by routed vessels has been estimated to be in excess of 7 hours. Since the saving of one hour's fuel will more than pay the cost of a shiprouting service, there is a valid economic argument for using such a service.

But it is not only in fuel savings that benefits are derived from shiprouting. An effective shiprouting service can also be a vital factor in ensuring that target arrival times are met and should also assist in reducing hull and cargo damage and loss. Insufficient information is available on cargo damage to allow the benefit of shiprouting to be quantified in this area, but there is enough information on hull damage to enable an assessment of the benefits of shiprouting in reducing hull damage to be made. Since cargo damage is much more common than hull damage at least the same level of benefit will apply, or indeed be much greater.

Miller (1983) examined the casualty rates of a large number of ships over several years and compared these rates for ships which were routed by Oceanroutes and those which were unrouted or routed by another agency. He found that for the sample studied - 150,000 North Atlantic and North Pacific voyages of which 26,000 were serviced by Oceanroutes - the casualty rates of unrouted vessels was 32 per cent higher than for routed vessels. Miller's work leads to the conclusion that in terms of hull damage there is a very persuasive case for subscribing to a shiprouting service.

In the offshore oil and gas industry it is perhaps more difficult to arrive at any definitive figures of the economic benefits of meteorological services. Perhaps Mason's comment that the customer may be reluctant to divulge the economic value has more than a grain of truth. Clearly the potential savings are very large; each day lost represents a significant sum of money wasted, and as in the provision of shiprouting services, a clear - if not easily quantifiable - case can be made for the application of meteorological services to all phases of offshore oil and gas operations.

References

Bacon, S. and Carter, D.J.T. (1991) "Wave Climate Changes in the North Atlantic and North Sea", International Journal of Climatology, Vol 11, 545-558.

Mason, B.J. (1966) "The role of meteorology in the national economy", Weather, 21: 382-393.

Tolstikov, B.I. (1968) "Benefits of meteorological services in the USSR" in Economic Benefits of Meteorology, WMO Bulletin 17: 181-186.

Maunder, W.J. (1972) "Assessing the value of weather information", unpublished paper, Wellington, New Zealand Meteorological Service.

Motte, R.J. and Calvert, S. (1991) Journal of Navigation, 41: 3.

Miller, James N. (1983) "Does Oceanrouting Reduce Casualties?", Oceanroutes Internal Report.

RISK MANAGEMENT IN THE OIL INDUSTRY INDEMNITIES AND INSURANCE

C.W. SPRAGUE,
Ince & Co.,
Knollys House,
11 Byward Street,
London, EC3R 5EN,
United Kingdom.

Abstract: In an ideal world risks can be avoided by careful analysis at the pre-contract stage of risks which may arise. However even in the best prepared and managed operations, accidents will happen. Risk management involves the apportionment of risks between the parties involved in contracts and effecting of insurance by those parties to cover the risks that they have assumed. For this to be effective, contracts must be carefully drafted.

1. Introduction

The obligations to insure that are imposed upon parties to an offshore contract depend very much upon how risks have been apportioned between them. This apportionment is done largely by way of indemnity clauses. The obligations as to insure and the clauses which embody these are intended to ensure that the parties, having assumed risks are then in a position to meet claims or losses which arise from them. Any consideration of insurance clauses must of necessity therefore include a consideration of the risks which are being insured against and the facts which determine how these risks are apportioned.

In an ideal world risks can be avoided. Careful analysis at the pre-contract stage of the risks that may arise will help towards avoidance of accidents. So too, at a later stage, will the carrying out of thorough site investigation, the use of adequate technology and equipment, proper safety procedures, adequate manning and effective supervision. In this way, it should be possible for example for an operator to ensure that a fixed platform, drilling equipment or in the case of a supply boat charterparty, a vessel, are adequate for their purpose. The Health and Safety at Work Executive hope that the Safety Case Procedure recommended by Lord Cullen in the Piper Alpha Report, and now being put into force, will largely eliminate risks arising from any inadequacies in the design, construction or operation of offshore installations.

However, the history of the offshore industry tells us that risks of this nature can be overlooked and that others, for example those due to unforeseeable external agencies beyond the control of the contractors and operators, human error, negligent workmanship or manufacture of equipment, cannot be avoided in even the best regulated of contracts.

Volume 30: Subsea International '93, 159–171.

2. Factors affecting the apportionment of risks

There are two important factors which may affect the way in which risks are apportioned. One of these is the state of the market at the time of the negotiations. When the offshore market is buoyant, a contractor will be able to negotiate a favourable contract. If the market is deflated, his position will be less strong.

The second factor affecting apportionment of risks will be the need, where possible, to avoid exposing a contractor, or for that matter a sub-contractor to risks which are out of all proportion to (a) the nature and cost of the service and/or equipment that they are supplying and to (b) their own financial resources. For example, a small supplier of relatively unimportant equipment or components, or fairly minor services, to the operator of an oil platform should not, in all fairness, have to bear the cost of substantial damage caused to the platform e.g. a blow-out, fire or pollution arising from the failure of the equipment or inadequacy of the services that it supplies. Such a contractor's involvement in terms of financial commitment to the operation as a whole will be minimal when compared with those of the operator and, leaving aside any questions of responsibility for the accident, it clearly makes sense for the operator, rather than the small supplier to which I have referred, to bear such risks or at least the greater part of such risks and effect the necessary insurance. This factor is important because it has to be considered whatever the state of the market and because it obviously affects the insurance provisions which have to be inserted.

Apportionment of risk will therefore depend in large part upon the relative size and resources of the parties negotiating a contract.

3. Identification of Risks

Most people concerned with offshore contracts will have their own check-lists of the risks which have to be considered. Obvious ones which spring to mind are, of course, damage to operators' and contractors' equipment or property and the consequences of bad weather in terms of damage to plant and equipment or delay in construction work or to the operation of an installation. In the context of indemnity clauses and consequent insurance provisions, these areas of risk can conveniently be identified as follows:

(a) Damage to or destruction of physical objects, such as the drilling unit or production platform itself and equipment belonging to the contractor or the operator, or sub-contractors.

(b) Liabilities created by statute or arising in contract or tort, including removal of debris, compensation to personnel for injury or loss of life, loss of effects and pollution.

(c) Expenses such as the cost or control of a well if there has been a blow-out or serious damage to an installation.

A typical set of indemnity provisions might be in the following terms:

9. Indemnities

9.1 Contractor's Indemnities

The contractor shall indemnify, hold harmless and defend the Company and its parent, subsidiary and affiliate corporations and Participants, and their respective officers, employees, agents and representatives from and against any claim, demand, cause of action, loss, expense or liability (including but not limited to the costs of litigation) arising by reason of:-

(a) Non-compliance with Laws

Claims by governmental authorities or others of any actual or asserted failure of the Contractor to comply with any law, ordinance, regulation, rule or order of any governmental or judicial body; and

(b) **Intellectual Property Infringement (Including Patents and Copyrights**

Actual or asserted infringement or improper appropriation or use by the Company, Participants or Contractor of trade secrets, proprietary information, know-how, copyright rights (both statutory and non-statutory), or patented or unpatented inventions or for actual or alleged unauthorised imitation of the work of others, arising out of the use of methods, processes, designs, information or other things originating with the Contractor, its employees, agents vendors or sub-contractors, and furnished or communicated to the Company by the Contractor or used by the Contractor in connection with performance of the Work; and

(c) **Injury to Contractor's Employees and Damage to Property of Contractor.**

Injury to or death of persons employed by or damage to or loss or destruction of property of the Contractor or its parent, subsidiary or affiliate corporations, or the Contractor's agents, sub-contractors or suppliers, irrespective of any contributory negligence, whether active or passive, of the party to be indemnified.

(d) **Third Party Injury and Property Damage**

Injury, death, or property damage, loss or destruction other than such as is described in Articles 9.1.(c) and 9.2 (a) and (b) up to a limit of U.S.$ 5,000,000.00 per occurrence, and arising directly or indirectly out of the acts or omissions of the Contractor or its sub-contractors, suppliers or their respective employees or agents, irrespective of any contributory negligence, whether active or passive, of the party to be indemnified.

(e) **Pollution**

Waste, debris, rubbish, liquid or non-liquid discharge or pollution of whatever nature which is dropped, seeped, discharged, spilled, blown out or leaked from equipment, apparatus, machinery, facilities or other property of the Contactor or its sub-contractors, suppliers, employees or agents, irrespective of any contributory negligence, whether active or passive, of the party to be indemnified.

9.2 **Company's Indemnities**

The Company shall indemnify, hold harmless and defend the Contractor and its parent, subsidiary and affiliate corporations, and their respective officers, employees, agents and representatives from and against any claim, demand, cause of action, loss, expense or liability (including but not limited to the cost of litigation) arising in relation to this Contract by reason of:-

(a) **Injury to Company's Employees and Damage to Property of Company**

Injury to or death of persons employed by or damage to or loss or destruction of property of the Company, Participants or their respective parent, subsidiary or affiliate corporations, irrespective of any contributory negligence, whether active or passive, of the party to be indemnified.

(b) **Pollution**

Waste, debris, rubbish, liquid or non-liquid discharge or pollution of whatever nature which is seeped, discharged, spilled, blown out or leaked from any underground reservoir or underwater pipeline or from any cargo barge, vessels or other equipment or facility of the Company, including fuel, lubricant or the like, irrespective of any contributory negligence, whether active or passive, of the party to be indemnified.

(c) **Intellectual Property Infringement (Including Patents and Copyrights)**

Actual or asserted infringement or improper appropriation or use by the Contractor, its subcontractors, Company or Participants of trade secrets, propriety information, know-how, copyright rights (both statutory and non-statutory), or patented or unpatented inventions or for actual or alleged unauthorised imitation of the work of others arising out of the use of methods, processes, designs, information other things originating with the Company, its employees, agents, vendors or contractors (other than Contractor) and furnished or communicated to the Contractor by the Company or used by the Company in connection with performance of the Work.

Several points that should be made about these indemnities.

First, the contractor and the operator each accept responsibility for, and agree to indemnify each other for death or injury to their respective employees and those of their sub-contractors and for damage to or loss of their own property, irrespective of any question as to the negligence of the party to be indemnified. This, as we shall see later, leads to a requirement that the operator and contractor should each insure its liability in respect of its own property and employees. This liability is unlimited and that is not entirely unreasonable. Even a relatively small contractor should be capable of insuring its own equipment and accepting liability for its own employees.

The second point worth noting is that the contractor accepts liability for damage to property of third parties (other than its own property and that of its sub-contractors) and in respect of death to or injury of third parties (other than its own employees and employees of its sub-contractors). The kind of liability with which we are concerned here includes injury to invitees on to the Platform or Rig, damage to third parties' property on, or in the vicinity of the Platform or Rig - for example the fracturing of pipelines during drilling operations or damage to vessels in the vicinity of the operation as a result of fire or explosion on the rig or platform. This liability is restricted to such claims which arise "directly or indirectly out of the acts or omissions of the contractors, or its sub-contractors, suppliers or their respective employees or agents." The limit of such liability is set at U.S.$5,000,000 per occurrence. The reasoning behind these restrictions is that contractors should not be made responsible for injury to third parties or damage to third party property where they have no control over the third parties' activities unless they (the contractors) are themselves responsible. The limit is imposed quite simply because the extent of exposure might be far greater than the contractor could itself meet or even insure against.

Thirdly it will be noted that responsibility for pollution is very carefully apportioned between the parties. Usually the operator will be responsible for pollution from the well, the reason behind this seems to be that there are certain types of damage, such as that to underground reservoirs and pollution caused by the well itself that are either impossible or extremely expensive to insure against. It is regarded as equitable that the party most able to control or avoid such damage, by site selection or selection of the rig or platform (i.e. the operator) should bear the ultimate risk. There are of course pollution risks which are more clearly within the contractor's control, particularly those relating to its own equipment and machinery and it is this type of risk which is covered by the contractor's indemnity in respect of pollution.

A fourth point which emerges from this clause is that the operator and contractor each provide, in the first paragraph of their indemnities, that they will indemnify, hold harmless and defend the other and its parent, subsidiary and affiliate corporations and participants, and their respective officers, employees, agents and representatives from and against any claim, demand, cause of action, loss, expense or liability arising from various risks. These risks are then listed. This is an introductory paragraph that is frequently used in indemnity clauses and owes its origin, I suspect, to American drafting. However there is in English law the doctrine of privity of contract which emerged in the 19th Century and which was set out very clearly in the leading case of Dunlop -v- Selfridge [1915] A.C. by Lord Haldane who stated:

> "Only a person who is a party to a contract can sue on it."

Apart from an exception relating to bills of lading, which need not concern us here, this remains good law. Unless therefore the parent companies, subsidiaries etc. of the operator or contractor are parties to the contract, they will not be able to enforce the indemnity against the operator or contractor in respect of loss or damage that they have suffered.

4. Indemnities and Insurance

It may seem that if a contract contains an indemnity clause of the type referred to above where the apportionment of responsibility is clear and does not depend upon any negligence on the part of either party, the party indemnified will not need to effect any insurance in respect of those risks or to consider insurance in relation to them at all. In other words, for the parties benefiting from them, the indemnity clauses act as a form of insurance against the risks referred to in them.

Strictly speaking of course, this is just what indemnity clauses should achieve. However, this will only be the case if they are correctly drafted and if there are other provisions in the contract which enable them to work effectively. These other provisions relate to insurance by the parties of the liabilities that they have assumed.

(a) Adequate insurance by the indemnifying party

An indemnity, however water-tight in itself, is only as good as the party giving it, or rather the insurance cover that the indemnifying party has effected. Usually, therefore, both the contractor and the operator will effect various insurance policies to cover their equipment and liabilities. The contract will oblige the contractors to effect specific insurance, for example, Workmen's Compensation and Employers' Liability Insurance, Public or Third Party Liability Insurance, Protection and Indemnity and Marine Hull Insurance (where the contractor is providing or operating vessels under the terms of the contract). The clause will often set out in some detail the scope of these insurances and also require that certificates of insurance be produced to the operators.

The insurance clauses which are complementary to the indemnity provisions discussed above are as follows:

> **10. Insurance**
>
> The Contractor shall, at its sole cost and expense, procure and maintain (and shall require its sub-contractors to maintain in effect) during the Contract Period, insurance coverage with insurers under forms and policies satisfactory to the Company as specified in Article 10.2 below. All such policies may be suitably endorsed as to territorial and/or navigational limitations to include the entire scope of operations contemplated by this Contract. Further, such policies other than Workers' Compensation, Employers Liability must include the Company and the Participants as additional insureds and shall provide that the policies will indemnify the additional insureds against claims brought by any other of the insureds.

Such policies other than Workers' Compensation and Employer's Liability must also provide that these insurances shall be primary and not contributing with any other insurance available to the Company or its Participants.

10.2 The following insurance coverage is to be maintained by the Contractor:-

(i) Employer's Liability and Workers' Compensation Insurance to comply with the statutory requirements.

(ii) Automobile Insurance to not less than the statutory requirements.

(iii) General Public Liability Insurance in respect of the persons and property of third parties to the extent of at least $5,000,000.00 for each incident arising out of the performance of the Work.

10.3 **Certificates of Insurance**

The Contractor shall furnish to the Company, prior to commencing work, Certificates of Insurance as evidence that policies providing the required insurance coverage are in full force and effect. Such certificates shall also provide that not less than thirty (30) days advance notice will be given in writing to the Company prior to cancellation, termination or material alteration of the said policies or insurance, and shall also contain a waiver of subrogation in favour of the Company, each of the Participants, and their respective officers, agents and employees.

There are several comments that should be made about these provisions.

First, it will be noted that the coverage which the contractor has to effect in respect of third party liability is limited to U.S.$5,000,000 which, of course, complements the limit under the indemnity in respect of third party liabilities.

Secondly, employers' liability coverage is to be obtained "to comply with the statutory requirements". Under English law (the Employer's Liability (Compulsory Insurance) Act 1969) an employer is required to insure and maintain insurance for £2,000,000 in respect of claims relating to any one or more of its employers arising out of any one occurrence. This statutory coverage may or may not be sufficient, depending upon the number of employees that a contract chooses to employ on a job at any one time and also upon the jurisdiction in which any proceedings might be brought. Awards by U.S. juries can of course be enormous by comparison with awards of English Courts in similar cases. It may well be that, having established the number of employees that a contract is likely to use at any one time in performing its duties, a specific level of employer's liability coverage should be stipulated over and above the statutory limit. It is in the contractors' interest to ensure that coverage is adequate but an operator cannot be sure, without specifying a limit, that the contractor has in fact protected its interest in this respect.

Thirdly, it will be seen that this particular wording (under clause 10.3) provides that certificates of insurance shall be supplied "... prior to commencing work". It is important that where possible a system should be set up in order to ensure that certificates are supplied. This particular clause however, is probably not satisfactory. An operator will wish to be certain that the requisite insurance is in place at all times and not just at the commencement of a job. In some cases the contractor's work will not last more than the duration of the certificate that is in force at the time that the contract is commenced. However in other cases the contract may be performed over a period of two or more years. The contract should, ideally, provide that certificates be tendered within a specified time of each insurance period.

Finally, these provisions contain references to "naming" i.e. the naming of the operator (and possibly associated companies) as an additional assured on the contractor's policies (clause 10.1) and "waiving" i.e. waiver of subrogation right (10.3). Naming and waiving is frequently provided for in contracts, often without too much thought being given as to whether it is either necessary or effective, and it is this process which I now propose to consider.

Although the above wording provides for naming and waiving, some contracts will provide for waiving where the assured is not named as a co-assured. For example:-

> "All insurance policies provided by contract under the contract (except policies which include operator as an additional named assured) shall be endorsed with the following wording to waive all express or implied terms of subrogation:-
>
> The insurers hereby waive their rights of subrogation against the operator (named) its affiliates and any officer, employee, agent, servant or contractor of the operator and its affiliates and against any individual, firm or corporation for whom or with whom the operator may be acting".

There will often be an identical clause dealing with the insurance policies effected by the operator.

It will be seen therefore that sometimes naming and waiving is used, but on other occasions waiving is provided for only where one party is not named as an additional assured on the other's policy.

Most of you will probably be aware of the principle of subrogation. This principle was very clearly set out in the leading case of Simpson -v- Thomson [1877] 3 App. Cas.279 (HL) where Lord Cairns L.C. said:-

> "I know of no foundation for the right of underwriters, except the well-known principle of law that where one person has agreed to indemnify another, he will, on making good the indemnity, be entitled to succeed to all the ways and means by which the person indemnified might have protected himself against or reimbursed himself for the loss."

The position is equally clearly stated, so far as marine insurance is concerned, in the Marine Insurance Act 1906 which states in section 79:

> (1) "Where the insurer pays for a total loss, either of the whole, or in the case of goods of any apportionable part of the subject matter insured, he thereupon becomes entitled to take over the interest of the assured in what ever may remain of the subject matter so paid for, and he is thereby subrogated to all rights and remedies of the assured in and in respect of that subject matter as from the time of the casualty causing the loss."

In short, the insurer, having indemnified the assured under a policy be it a policy on property or a liability policy, can then "stand in the shoes of the assured" and pursue any remedies against third parties for tort or breaches of contract which the assured may have. For example, let us assume that a contractor is sued by an employee of another company in respect of injury suffered on the Platform. The contractor is liable but, there is an arguable case for saying that the operator was also partly to blame. Having paid the claim under the contractor's public liability policy, the underwriter of that policy would be subrogated to the contractor's rights against the operator.

How would an indemnity in the contract of the kind that we have considered affect the situation? You will recall that in the clause, the contractor had to indemnify the operator for injuries to third parties "arising directly or indirectly out of the acts or omissions of the contractor ... irrespective of any contributory negligence" on the part of party to be indemnified. The operator would be perfectly entitled to say that the claim related to a risk in respect of which the contractor had agreed to provide an indemnity. It is a basic principle of subrogation that an insurer, when suing in the name of the assured, will have no better rights than the assured. Assuming therefore that the indemnity was in the form alleged by the operator, it would provide a good defence to such a subrogation claim.

This sort of situation inevitably raises the question as to why, where the indemnity clauses are in this form, a waiver of subrogation is necessary at all. So far as the United States is concerned, certain types of indemnity clause, for example those involving one party undertaking to meet personal injury or loss of life claims irrespective of whether these arose as the result of the negligence of the other party, may be unenforceable on the ground that they are contrary to public policy. This will not be the case so far as federal law is concerned and it is this law which is applied in admiralty matters. However, many states have laws similar to the English Continental Shelf and Oil and Gas Acts. These extend the jurisdiction of a particular state's Courts so as to include the sea bed and anything attached to it, within a specified area adjacent to their coastline. This would include fixed equipment such as drilling platforms. The State of Louisiana for example has passed an act which stipulates that this type of indemnity is unenforceable and it is highly probable that where personal injury or death occurs on a fixed platform within the jurisdiction of the Louisiana Courts this law, rather than the less stringent (in this respect) U.S. Federal Law, would apply.

This problem is less likely to arise in England. While Section 2(1) of the Unfair Contract Terms Act 1977 stipulates that a party cannot in any circumstances exclude or restrict his liability for death or personal injury resulting from negligence, the application of the Act to indemnity clauses is very limited. Section 4 states that where the party giving the indemnity is a consumer within the meaning of the Act, the indemnity will only be enforceable if it is reasonable. However, a consumer is a party that does not make a contract in the course of business. No party to an offshore contract is likely to fall into this category and it is not therefore necessary to consider the vexed question as to whether the indemnity is reasonable.

There are, however, several reasons why under English law a waiver or subrogation in an offshore contract may be necessary. First, the granting of an indemnity may, as I have already indicated, remove the right of the underwriters to pursue an action by way of subrogation. In Castellain -v- Preston [1883] 11 QBD 380 Brett L.J. said:

> "As between the underwriter and the assured, the underwriter is entitled to take advantage of every right of the assured, whether such right consists in contract, fulfilled or unfulfilled or in remedy for tort capable of being insisted on ...".

At the inception of cover, there is an obligation on the part of the assured to make full disclosure of all material facts to underwriters. The best definition of the term "material" is to be found in section 18 of the Marine Insurance Act which states:

> "Every circumstance is material which would influence the Judgment of a prudent insurer in fixing the premium, or determining whether or not he will take the risk."

This test has also been applied to the obligations of an assured under a non-marine insurance policy (see March Cabaret -v- London Assurance [1975] 1 LLR. 169). An indemnity clause which removes the underwriters' rights to subrogation would clearly be a material fact and a provision in a contract which requires that there be a waiver of subrogation clause in an insurance policy will ensure that the underwriter is made aware that he is losing his right of subrogation and that the insurance is not in any way vitiated.

Another reason why a waiver of subrogation clause may be desirable is that notwithstanding the intention of the parties as to the apportionment of risk, the indemnity clause, possibly because of inadequate drafting, may not have the required effect. In other words the underwriter might, but for a waiver of subrogation clause, be able to pursue a claim against the other party to the contract for whom the indemnity clause was intended to provide protection.

Finally, one of the purposes of an insurance package, quite apart from reinforcing, and being consistent if possible, with the indemnity provisions in the contract is to remove the possibility of arguments between the parties as to which should bear any particular risk. Provided that the insurer of one or other of the parties has met a claim for property damage or liability to a third party, there should not be arguments between the parties to the contractors as to which one is ultimately responsible. A waiver of subrogation clause will reduce the chances of such conflicts arising between the parties to the contract.

(c) Naming of Additional Assured

We have seen that it is very common for insurance clauses in a contract to provide that one party, for example the operator, shall be named as an additional assured on the policies effected by the other party, the contractor. As I have indicated already, often, although not always, where one party to a contract is named on an insurance policy effected by another there will also be a waiver of subrogation. The procedure is commonly known as "naming and waiving".

The requirement as to naming on policies will often be across the board and will not differentiate between various types of policy.

There are some situations where naming is positively beneficial. For example it will be in an operators' interest to be named as an additional assured in a contractors' insurance policy which protects against liability to third parties (since that liability might arise not only out of the contractors' activities but as a result of those of the operator as well). Having an insurer who is obliged to bear the cost of investigation in defence is better than having an underwriter who will merely, at the end of the day, pay a sum awarded under a Judgment. A few years ago I was concerned on behalf of a rig operator in a case involving the death of three riggers. They were not employees of the contractor. Nevertheless, my Clients were entitled to an indemnity from the contractor and this was accepted by the contractor very shortly after both parties had completed their investigation of the facts. The dependants of the deceased sued both the contractors and the operator and, for some reason best known to themselves, the Solicitors representing the contractors' insurers were not prepared to take over the conduct of my Clients' case for some considerable time. In the end they did so and the costs that we had incurred were included in the indemnity. However, had my Clients been named as an additional assured, this problem would not have occurred and indeed it is likely that both the initial investigation and the defence of the claim would have been handled jointly, probably by one firm of solicitors.

However a joinder as an additional assured is not always so desirable or satisfactory. Clearly there is little point in naming an operator as an additional assured on the contractors' employers liability policy since the operator would have not interest in the policy covering the contractors' liability to its employees except in the case of a claim made by someone who is prima facie an employee of the contractor but is in fact a borrowed employee of the operator.

Different interests of the contractual parties can create further problems where both are named on the same policy. One situation in which such a problem might occur is where a contractor under an offshore contract is operating a supply vessel. He has, inter alia, protection and indemnity coverage in respect of this vessel. When the vessel is alongside the operators' platform, oil is spilled on to the deck as the result of the operators' negligence. A seaman on the contractor's vessel slips on the oil and is injured. He sues the operator who is named as an additional assured on the contractor's policy. When however the operator seeks to recover under the policy, the underwriters might well adopt the view that the policy answers to claims in respect of liability arising from the operation of the vessel whereas this incident arose as the result of negligent operation of the platform. Accordingly, there can be no cover in respect of the operators' liability to the seamen.

A similar problem arose in the case of Petrofina -v- Magnaload [1983] 2 LLR. 91. Here a gantry collapsed during construction work at an oil refinery. There was extensive damage to the contract works and two workmen were killed. The owners of the refinery made a claim under the Contractors' All Risks Policy and, having paid the claim, the insurers sought by way of subrogation to pursue the owners and operators of the heavy lifting equipment involved in the accident. The Defendants contended that since they were covered under the policy, they could not be sued. The policy provided:

> "The insurers will indemnify the insured against loss of or damage to the insured property whilst at the contract site from any cause not hereinafter excluded occurring during the period of insurance."

The "insured" were defined in the schedule as "the owners of the refinery and/or main contractors and/or sub-contractors". The Court had to decide two questions. First, whether the defendants actually fell within the definition of the "insured" under the policy and secondly whether the insured were covered in respect only of their own property or the property for which they were responsible or whether their cover extended to their liability in respect to the whole of the contract works.

The Judge decided that, on the facts, the defendants were sub-contractors and were therefore insured under the policy. He also held that each of the named assured, including the sub-contractors, was insured in respect of the whole of the contract works. Accordingly, the contractors all risks insurers were not allowed to pursue the subrogation claim against the sub-contractors.

I return now to the inclusion of a waiver of subrogation clause where one party is named as an assured on the other party's policy. Again, the reasons for including such a clause in the policy are not as straightforward as they might at first seem. It was accepted by the Judge in the Petrofina case (above) that an underwriter could not exercise rights of subrogation in the name of one co-insured against another co-insured under the same policy. In those circumstances, a waiver of subrogation clause was seen to be unnecessary. However, this assumes that the interests of the co-insured are identical and that both parties are entitled to recovery in respect of the accident in question. In the case of the oil spilling on to the deck of the supply vessel, or if in the Petrofina case the decision had gone against the defendants, a waiver of subrogation clause would have been necessary. It is to cover such eventualities that clauses of this kind are incorporated.

However care must be taken when naming additional assured. First, where an operator requires that it shall be named on a contractor's insurance policies, the operator itself will usually have its own insurance. Steps must be taken to ensure that there is no double insurance. This explains the inclusion in clause 10.1 of the insurance clauses that we discussed earlier of a provision that the contractors' insurances shall be stated to be primary and not contributing with any other insurance available to the operator.

The naming of the co-assured must also be clear. This was highlighted in a recent case Appledore Fergusson Shipbuilders -v- Stone Vickers which was decided in the Court of Appeal this summer. Stone Vickers had supplied a single screw controllable pitch propeller to Appledore Fergusson. They then had to sue for the contract price. Appledore counterclaimed alleging that the propeller was improperly manufactured and, in technical jargon, "sang". Stone Vickers' defence was to the effect that Appledore had claimed under their insurance policy and that this was a subrogated claim by Underwriters and must fail because Stone Vickers were a co-assured and because of the "no recourse" i.e. waiver of subrogation clause in the policy. This defence failed because the insurance documents did not make it sufficiently clear that Stone Vickers were indeed to be insured or that the no recourse clause applied to them. They were not specifically named in the relevant declaration under the policy and the contract with them had not been finalised when the declaration was made. This kind of problem can be avoided if the additional parties are clearly identified and intentions are clearly defined in the policy.

To summarise, great care is needed when naming a party as an additional assured under a policy just as it is when drafting indemnity clauses. When a waiver of subrogation and the naming of an additional assured is necessary, clear words must be used. One oil company's drilling contract that I have seen stated merely "any insurance taken out by contractors shall be to the benefit of contractor and company and shall be endorsed to provide that the underwriters waive their rights of recourse on company." "Benefit" meant, of course, naming and "recourse" meant subrogation, but these terms did not clearly indicate the obligations of the parties.

Very careful thought should be given to the drafting of indemnities and of insurance clauses to which are intended to complement them. Careful drafting can avoid a good deal of difficulty and may even do lawyers out of a job!

Zeitfracht Medien GmbH
Ferdinand-Jühlke-Straße 7
99095 Erfurt, Deutschland
produktsicherheit@kolibri360.de